IX International Conference Days of Applied Mathematics (IX ICDAM)

Journal of Physics: Conference Series Volume 2515

San Jose de Cucuta, Colombia
20 – 23 September 2022

ISBN: 978-1-7138-7468-3
ISSN: 1742-6588

Printed from e-media with permission by:

Curran Associates, Inc.
57 Morehouse Lane
Red Hook, NY 12571

Institute of Physics
Dirac House, Temple Back
Bristol BS1 6BE UK

Phone: 44 1 17 929 7481
Fax: 44 1 17 920 0979

techtracking@iop.org

Additional copies of this publication are available from:

Curran Associates, Inc.
57 Morehouse Lane
Red Hook, NY 12571 USA
Phone: 845-758-0400
Fax: 845-758-2633
Email: curran@proceedings.com
Web: www.proceedings.com

TABLE OF CONTENTS

Journal of Physics: Conference Series 2515 (2023) 011001 doi:10.1088/1742-6596/2515/1/011001

IX International Conference Days of Applied Mathematics

O Valbuena[1], E Gelvez-Almeida[1], and E D V-Niño[2]

[1] Facultad de Ciencias Básicas y Biomédicas, Universidad Simón Bolívar, San José de Cúcuta, Colombia

[2] Materials Science and Technology Research Group, Foundation of Researchers in Science and Technology of Materials, Bucaramanga, Colombia

E-mail: info@foristom.org, oscar.valbuena@unisimon.edu.co

These proceedings are the academic and scientific contributions of the IX International Conference / Days of Applied Mathematics (ICDAM), carried out in San José de Cúcuta, Colombia, from September 21-23, 2022. The main objective of the IX ICDAM is to share concrete experiences of mathematical and physics knowledge, as well as make contributions to problem-solving in different contexts of science and engineering; likewise, the IX ICDAM highlights contributions recent that participants are making to the applied mathematics and scientific computing community. The IX ICDAM was organized by the "Facultad de Ciencias Básicas y Biomédicas" of the Universidad Simón Bolívar, San José de Cúcuta, Colombia. The IX ICDAM research works were presented by students, professors, and researchers from Chile, Argentina, Mexico, Dominican Republic, Perú, Venezuela, and Colombia, through hybrid participation (on-site and virtual).

The topics developed in the IX ICDAM were the following: (i) computer science in mathematics and engineering, (ii) medical physics, (iii) mathematical physics, (iv) modeling, simulation, and diagnostics, and (v) education and communication in science and engineering. The ICDAM is part of the training and research processes of the academic program of mathematics and computer science at Universidad Simón Bolivar, framed within the high quality of education and the relevance of a program aligned with the new times in the digital era and the modeling of scientific knowledge in areas as diverse as mathematics, physics, biology, chemistry and the applications in medicine, engineering, management, business, education, and others.

The ICDAM offers researchers in training space for participation and training through interactions and dialogues with consolidated researchers in the different areas of knowledge that participate in the framework of the conferences. It also allows students, teachers, and researchers to attend a space for updating and knowledge appropriation from the workshops presented at the sections. It is important to highlight the visibility that ICDAM has recently had due to the participation of masters and doctoral students, from different countries of Latin America, in that they had presented the progress of the graduate work; moreover, the bonds of friendship among the participants have been consolidated, which has even led to the realization of collaborative works between authors from the participating institutions.

The international reputation of ICDAM and Universidad Simón Bolivar has increased by to the participation of students, professors, and researchers from different countries; in this sense, the Universidad Simón Bolivar has positioned itself as one of the recognized private Colombian institutions with the participation high of researchers in applied mathematics and physics, scientific modeling, and medical image processing. Likewise, the ICDAM has made the region of Norte de Santander, Colombia, visible as a reference for academic and scientific mobility, making academics, scientific, cultural, gastronomic, sports, and social inclusion aspects known to the international community.

The organizing and editorial committee is grateful for the unconditional support of professor Carlos Corredor Pereira (r.i.p.); likewise, thanks to the following partner institutions: Institute of Physics (IOP) Publishing, Foundation of Researchers in Science and Technology of Materials (FORISTOM), Society for Industrial and Applied Mathematics, Colombian Section (CoSIAM), Asociación Colombiana de Facultades de Ciencias (ACOFACIEN), Sociedad Colombiana de Matemáticas, and the Laboratorio de Investigaciones Tecnológicas en Reconocimiento de Patrones (LITRP), Mathematics, and Computer Science program of the Universidad Simón Bolívar, and the especially extend their gratitude to all the authors and participants for the valuable contributions to this publication; as well as to the reviewers for all the recommendations and constructive criticism that helped to improve the works presented. Therefore, we hope you enjoy reading this volume of the Journal of Physics: Conference Series (JPCS), IOP Publishing, and that these manuscripts contribute to the generation and transfer of knowledge, as well as to the strengthening of research in Colombia.

Finally, the editor hopes that those interested in the area of physical and mathematical applications can enjoy this reading, of the volume of the JPCS, which reflects a wide variety of current issues.

"Education and expression freedom for anyone".

Journal of Physics: Conference Series **2515** (2023) 011001 doi:10.1088/1742-6596/2515/1/011001

Organizing Committee

- Oscar Valbuena Prada (Chairman)
- Elkin Gelvez Almeida (Co-chairman)
- Juan Pablo Salazar Torres (Co-chairman)

National Scientific Committee

- Ph.D. Diana Haidive Bueno Carreño
  Universidad Javeriana, Santiago de Cali
- Ph.D. Adolfo Pimienta Pimienta
  Universidad Simón Bolívar, Barranquilla
- Ph.D. José José Barba Ortega
  Universidad Nacional de Colombia, Bogotá
- Ph.D. Johan Manuel Redondo
  Universidad Católica de Colombia, Bogotá
- Ph.D. Ely Dannier V. Niño
  Fundación of Researchers in Science and Technology of Materials, Bucaramanga
- Ph.D. Fredy Humberto Vera Rivera
  Universidad Francisco de Paula Santander, San José de Cúcuta
- Ph.D. Marcela Flórez Romero
  Universidad Simón Bolívar, San José de Cúcuta
- Ph.D. Miguel Ángel Vera
  Universidad Simón Bolívar, San José de Cúcuta
- Ph.D. Antonio José Bravo
  Universidad Simón Bolívar, San José de Cúcuta

International Scientific Committee

- Ph.D. Oscar Carrillo
  CPE Lyon, France
- Ph.D. Santiago Arceo Díaz
  Instituto Tecnológico de Colima, Mexico
- Ph.D. Karina Alejandra Vilches Ponce
  Universidad Católica de Maule, Chile
- Ph.D. Marco Mora
  Universidad Católica de Maule, Chile
- Ph.D. Ricardo Javier Barrientos Rojel
  Universidad Católica del Maule, Chile
- Ph.D. Primitivo Belén Acosta-Humánez
  Instituto Superior de Formación Docente Salomé Ureña, Dominican Republic
- Ph.D. Maura Leonor Vásquez de Ramírez
  Universidad Central de Venezuela, Venezuela
- Ph.D. Guillermo Ramírez
  Universidad Central de Venezuela, Venezuela

Sponsor

- Universidad Simón Bolívar

Partners

- IOP Publishing
- Sociedad Colombiana de Matemáticas
- Asociación Colombiana de Facultades de Ciencias
- Fundación of Researchers in Science and Technology of Materials
- Society for Industrial and Applied Mathematics, Colombian Section
- Laboratorio de Investigaciones Tecnológicas en Reconocimiento de Patrones

UNIVERSIDAD SIMÓN BOLÍVAR
BARRANQUILLA Y CÚCUTA - COLOMBIA | VIGILADA MINEDUCACIÓN

Foristom Foundation of Researchers in Science and Technology of Materials

LITRP Laboratorio de Investigaciones Tecnológicas en Reconocimiento de Patrones

ACOFACIEN

siam COLOMBIA

Sociedad Colombiana de Matemáticas

Journal of Physics: Conference Series **2515** (2023) 011001 doi:10.1088/1742-6596/2515/1/011001

Journal of Physics: Conference Series **2515** (2023) 011002 doi:10.1088/1742-6596/2515/1/011002

Peer Review Statement

All papers published in this volume have been reviewed through processes administered by the Editors. Reviews were conducted by expert referees to the professional and scientific standards expected of a proceedings journal published by IOP Publishing.

- **Type of peer review:** Double Anonymous
- **Conference submission management system:** By email received from the organizing and editorial committee of the conference and Morressier web platform; the correspondence authors make the submission by Morressier web platform and email.
- **Number of submissions received:** 17
- **Number of submissions sent for review:** 14
- **Number of submissions accepted:** 14
- **Acceptance Rate (Submissions Accepted / Submissions Received × 100):** 82.4
- **Average number of reviews per paper:** 2
- **Total number of reviewers involved:** 10
- **Contact person for queries:**
  Name: Ely Dannier V. Niño
  Email: info@foristom.org
  Affiliation: Foundation of Researchers in Science and Technology of Materials (FORISTOM)

Journal of Physics: Conference Series **2515** (2023) 012001 doi:10.1088/1742-6596/2515/1/012001

Existence of solutions for an optimal control problem in forestry management

A Altamirano-Fernández[1], A Rojas-Palma[1], and S Espinoza-Meza[2]

[1] Doctorado en Modelamiento Matemático Aplicado, Universidad Católica del Maule, Talca, Chile

[2] Facultad de Ciencias Agrarias y Forestales, Universidad Católica del Maule, Talca, Chile

E-mail: alex.altamirano@alu.ucm.cl

Abstract. Physical, biological, and economic factors remain inexhaustible sources when determining silvicultural management strategies that maximize the net benefit of timber volume. The objective of this paper is to formulate an optimal control problem to maximize the net benefit of timber volume and to demonstrate the existence of optimal solutions to this problem. The optimal control theory will be used, through the formulation of a bioeconomic optimal control problem subjected to a system of three ordinary differential equations that describe the interactions between the living biomass, the intrinsic growth of the biomass, and the burned area. Four control variables are associated with these differential equations, which are reforestation, felling, thinning, and fire prevention. Assuming adequate conditions on the control variables, the existence of optimal solutions to the optimal control problem is obtained using Filippov's theorem. In conclusion, the optimal control problem is well formulated and solutions exist. This will allow us to define forest management strategies that optimize the economic gain of the process, taking into consideration the physical, biological, and economic phenomena involved in the dynamics of forest plantations.

1. Introduction

The extraction rate of forest plantations is a social and economic decision within certain biological and physical constraints [1, 2]. Plantation forestry for industrial purposes was not a priority until the mid-20th century in countries such as Australia, New Zealand, the United States, and Chile [3]. However, large-scale forest management programs were initiated in many countries in the 1960s, and the early part of this century experienced a significant expansion in the area of fast-growing exotic forest plantations for industrial purposes, at a rate of 4.5 million hectares per year [4], applying a variety of economically beneficial forestry strategies [5].

The major part of forest management work is the maximization of income from timber resources, but with several ecological constraints on the time of felling (i.e., the rotation age) [6]. According to [7], they studied that optimal rotation age is a crucial factor in forest management strategies in both for timber production and carbon capture. In [8], they present a Faustmann model that determines the optimal rotation age for felling, in both economic and biological aspects. They determined that economic rotations were longer than biological rotations by almost 2 years, at a discount rate of 0.005. Also in [9, 10], they presented regression models for the optimal rotation age applied to the species *Pinus radiata* under fire risk. They obtained that when there is a high probability of fire, the rotation age is shortened (from 21 to 32 years) and if the fire risk is low, the rotation age is prolonged (from 53 to 64 years).

On the other hand, [11, 12], applied the optimal control theory in palm plantations to maximize oil production and carbon sequestration while maintaining biomass at a certain level. Recently [13] formulated a mathematical model that studies the dynamics of carbon capture in forest plantations, which was taken as a reference for the present work. In this context, the main objective of this paper is to formulate an optimal control problem that maximizes the net benefit of timber volume for fast-growing forest plantations and to demonstrate the existence of solutions. In this work we contribute from a mathematical point of view to the comprehension of phenomenon that obeys physical laws such as the fire propagation and the burned area. The model addresses a deterministic dynamic optimization problem, given by three state variables; live biomass, intrinsic biomass growth, and burned area. Four control variables are incorporated into these variables: reforestation, felling, thinning, and fire prevention. Filippov theorem is used for demonstration existence of solutions that guarantee the existence of an optimizing optimal control quadruple.

2. Mathematical model

There are a variety of mathematical models in the literature that study the optimal rotation age in forest plantations (*e.g.*, [14–16]). For the present work, we will follow the same ideas of the mathematical model presented in [13], which consists of a system of three ordinary differential equations (ODEs) involving three state variables the live biomass $B(t)$, the intrinsic biomass growth $r(t)$ and the burned area $I(t)$. Four control variables were associated with reforestation $R(t)$, felling $F(t)$, fire prevention $S(t)$, and thinning $T(t)$. The notation "point" represents the derivative of a variable with respect to time t, in the dynamic system given by Equation (1).

$$\begin{cases} \dot{B}(t) = r(t)B(t)\left(1 - \frac{B(t)}{K}\right) + [\beta R(t)]B(t) - [\mu_1 I(t) + \sigma F(t) + \tau T(t)]B(t), \\ \dot{r}(t) = r_0 - \rho r(t) + \nu T(t), \\ \dot{I}(t) = \mu_2 I(t)\left(\frac{B(t)}{1+B(t)}\right) - \theta S - \eta T(t) - hI(t). \end{cases} \tag{1}$$

In Equation (1), satisfying the initial conditions $B(t_0) = B_0, r(t_0) = r_0$ and $I(t_0) = I_0$; their notations, definitions, and units for the variables are described in Table 1.

Table 1. Notation, definition, and units of each variable.

Notation	Definition	Unit
$B(t)$	Volume of living biomass	m^3 ha^{-1}
$r(t)$	Intrinsic growth of biomass	yr^{-1}
$I(t)$	Burned area	m^2 yr^{-1}
$R(t)$	Forest reforestation	ha yr^{-1}
$F(t)$	Forest felling	ha yr^{-1}
$S(t)$	Fire prevention	US$ m^2 ha^{-1} yr^{-1}
$T(t)$	Forest thinning	ha yr^{-1}

Where, in Table 1, the variables $B(t) \geq 0$, $r(t) \geq 0, I(t) \geq 0, R(t) \geq 0, F(t) \geq 0$, $S(t) \geq 0$ and $T(t) \geq 0$, the parameters $(K, \sigma, h, \mu_1, \mu_2, \tau, \beta, r_0, \rho, \eta, \theta, \nu) \in \mathbb{R}^{12}_{+}$. Where the biomass dynamics has a logistic growth with a carrying capacity of K, β is the rate of increase of biomass concerning reforestation. In this model biomass decreases due to the effects of fire, felling, and thinning, where μ_1 is the rate of biomass decrease due to fire, σ is the rate of biomass decrease due to felling and τ is the instantaneous rate of biomass decrease due to thinning. The dynamics of intrinsic growth are given by: r_0 is the maximum growth rate (under ambient conditions) [17], ρ is the natural mortality rate with respect to individual growth, and ν is the rate of increase by the long-term effect of thinning which indirectly affects the individual growth.

Finally, the burned area is directly proportional to the amount of biomass. For this reason, an inhibition term $I(t)\left(\frac{B(t)}{1+B(t)}\right)$ is added that makes the area burned not grow in an unlimited way. Here μ_2 represents the fire rate, θ is the fire prevention rate, η is the thinning rate and h is the relative humidity threshold at the time the fires occur.

2.1. Optimal control problem

In this section, we consider a bioeconomic optimal control problem of Equation (1) to determine the optimal management strategies to maximize the net benefit of timber volume. Based on the ideas of [18]. In Equation (2), we consider the bioeconomic problem where the objective functional J represents the present value of the future flow of net benefit associated with the process of harvesting the forest resource.

$$J(F,T,R,S) = \int_0^{T_f} U(t)e^{-\delta t}dt, \tag{2}$$

where U is the utility function, which is defined as follows in Equation (3).

$$U(t) = P_1(t)B(t) + P_2(t)B(t) - C_1(t)F(t) - C_2(t)T(t) - C_3(t)R(t) - C_4(t)S(t). \tag{3}$$

From the Equation (3), it is known that P_1, P_2 are the sales value of the timber obtained from thinning and final felling; while C_1, C_2, C_3, and C_4 are the costs associated with the control variables, each of these is time-dependent parameters and are written in Equation (4) as follows.

$$P_1(t) = p_1F(t),\ P_2(t) = p_2T(t),\ C_1(t) = c_1F(t),\ C_2(t) = c_2T(t),\ C_3(t) = c_3R(t),\ C_4(t) = c_4S(t). \tag{4}$$

Where p_1 is the constant price per unit of biomass of the felled stock, p_2 is the constant price per unit of biomass of the thinned stock, and the parameters c_1, c_2, c_3 and c_4 are the constant costs associated with the rate of felling, thinning, reforestation, and fire prevention, respectively. Therefore, taking into account Equation (4) and replacing in Equation (2), the functional J is expressed in Equation (5) as follows.

$$J(F,T,R,S) = \int_0^{T_f} [p_1F(t)B(t) + p_2T(t)B(t) - c_1F(t)^2 - c_2T(t)^2 - c_3R(t)^2 - c_4S(t)^2]e^{-\delta t}dt, \tag{5}$$

subject to the system state variables Equation (1), the final time T_f is fixed and the state variables B, r, and I are assumed free at the final time, as shown in Equation (6).

$$\begin{aligned} &B(0) = B_0, \quad r(0) = r_0, \quad I(0) = I_0, \\ &B(T_f) = \text{free}, \quad r(T_f) = \text{free}, \quad I(T_f) = \text{free}. \end{aligned} \tag{6}$$

The coefficient $e^{-\delta t}$ is the discount factor over time, where δ is the interest rate. However, using quadratic controls allows for deriving an analytical expression for the optimal yield. The resulting solution is different from the bang-bang solution that is usually obtained in the case of a linear function [19].

2.2. Existence of the control problem

The existence of the optimal control quadruple can be obtained using the studies by [20–22]; from the Equation (1), we denote that the control variables $\mathbf{u}(t) = (R(t), F(t), T(t), S(t)) \in \mathcal{U}$ which are associated with the state variables $\mathrm{X}(t) = (B(t), r(t), I(t))$, are bounded and measurable according to Lebesgue, as shown in Equation (7).

$$\mathcal{U} = \{\mathbf{u} : F \in [0, \hat{F}], R \in [0, \hat{R}], T \in [0, \hat{T}], S \in [0, \hat{S}], t \in [0, T_f]\}, \tag{7}$$

where $\mathcal{U}$ is the class of admissible controls. More precisely, the following theorem is stated.

Theorem 1. If $c_1 > \frac{p_1}{2}$ and $c_2 > \frac{p_2}{2}$, there exists an optimal control quadruple $(F^*, T^*, R^*, S^*) \in \mathcal{U}$, such that it satisfies in Equation (8).

$$J(F^*, T^*, R^*, R^*) = [(F, T, R, S) \in (\mathcal{U})]\max(J(F, T, R, S)). \tag{8}$$

Subject to the Equation (1) with non-negative initial conditions; if the following [22] conditions are satisfied:

(i) The set of controls and state variables of the problem is not empty.
(ii) The admissible control set $\mathcal{U}$ is bounded and convex.
(iii) The right-hand side of Equation (1) is bounded by a nonlinear function that depends on the state and control variables.
(iv) The integrand of the objective function is convex in the control set $\mathcal{U}$.
(v) There exist positive constants $d_1, d_2 > 0$ and $d_3 > 1$ that satisfy the integrand of the objective function, Equation (9).

$$J(R, F, S, T) \leq d_2 - d_1(|F|^2 + |T|^2 + |R|^2 + |S|^2)^{d_3/2}. \tag{9}$$

Proof. (i) It is known that solutions of Equation (1) are bounded in a finite time interval, making use of a result of [23], the existence of a solution for the controlled system can be assured. (ii) It is observed that the set of admissible controls $\mathcal{U}$, topologically is closed and convex by definition. (iii) The control system given by Equation (1), is rewritten in Equation (10) in a matrix form.

$$\dot{\mathrm{X}}(t) = A\mathrm{X} + \mathcal{F}(\mathrm{X}) + Y(\mathbf{u}), \tag{10}$$

where the terms of Equation (10) are defined in Equation (11).

$$\mathrm{X} = \begin{pmatrix} B \\ r \\ I \end{pmatrix}, \quad A = \begin{pmatrix} (\beta R) - (\sigma F + \tau T) & 0 & 0 \\ 0 & -\rho & 0 \\ 0 & 0 & h \end{pmatrix},$$

$$\mathcal{F}(\mathrm{X}) = \begin{pmatrix} rB\left(1 - \frac{B}{K}\right) - \mu_1 IB \\ 0 \\ \mu_2 I\left(\frac{B}{1+B}\right) \end{pmatrix}, \quad Y(\mathbf{u}) = \begin{pmatrix} 0 \\ r_0 + \nu T \\ -\theta S - \eta T \end{pmatrix}, \tag{11}$$

the derivative with respect to the time t is represented by $\dot{\mathrm{X}}$; the Equation (10) is a nonlinear system with a bounded coefficient. Then we can define Equation (10) as follows in Equation (12).

$$G(\mathrm{X}) = A\mathrm{X} + \mathcal{F}(\mathrm{X}) + Y(\mathbf{u}). \tag{12}$$

From Equation (12), we consider the second term and apply the Hölder inequality and absolute value, then we have Equation (13).

$$\begin{aligned} |\mathcal{F}(\mathrm{X}_1) - \mathcal{F}(\mathrm{X}_2)| =& |(r_1 B_1 (1 - \frac{B_1}{K}) - \mu_1 I_1 B_1) - (r_2 B_2 (1 - \frac{B_2}{K}) \\ & - \mu_1 I_2 B_2) + (\mu_2 I_1 (\tfrac{B_1}{1+B_1})) - (\mu_2 I_2 (\tfrac{B_2}{1+B_2}))| \\ & \leq |r_1 B_1 - r_2 B_2| + |\tfrac{r_2 B_2^2}{K} - \tfrac{r_1 B_1^2}{K}| + \mu_1 |I_1 B_1 - I_2 B_2| + \mu_2 |I_1 B_1 - I_2 B_2| \\ & \leq M_1 |B_1 - B_2| + M_2 |r_1 - r_2| + M_3 |I_1 - I_2| \\ & \leq \max(M_1, M_2, M_3)(|B_1 - B_2| + |r_1 - r_2| + |I_1 - I_2|), \end{aligned} \tag{13}$$

where, $M_1 = \left[r_m + (\mu_1 + \mu_2)\left(\frac{B_m(r_m+\beta R)}{\mu_1(1+B_m)} + r_m\right) + \frac{r_m}{K}(2B_m)\right]$, $M_2 = \left(B_m + \frac{B_m^2}{K}\right)$ and $M_3 = B_m(\mu_1 - \mu_2)$. In addition, it is known from the bounding of the solutions of Equation (1), $r_m = \frac{r_0+\nu T}{\rho}$ and $B_m = K + K\frac{\beta R}{r_m}$. In Equation (14), the constant M is positive, taking $M = \max(M_1, M_2, M_3, \|A\|) < \infty$ which is independent of the state variables, we have Equation (14).

$$|G(\mathrm{X}_1) - G(\mathrm{X}_2)| \leq M|\mathrm{X}_1 - \mathrm{X}_2|. \tag{14}$$

Therefore, the function $G(\mathrm{X})$ is said to be uniform Lipschitz continuous. From the definition of $\mathcal{U}$ and the constraint of B, r and I we can assure the existence of the solution of the Equation (1) [24, 25].

(iv) To show the concavity of the integrand of the objective function, let us denote as follows in Equation (15).

$$\mathcal{N}(t, X, u) = [p_1 F(t) B(t) + p_2 T(t) B(t) - c_1 F(t)^2 - c_2 T(t)^2 - c_3 R(t)^2 - c_4 S(t)^2]. \tag{15}$$

For this, Equation (16) presents the definition of a convex set.

$$(1-q)\mathcal{N}(t, X, u) + q\mathcal{N}(t, X, v) \leq \mathcal{N}(t, X, (1-q)u + qv), \tag{16}$$

let u, v are two control vectors and $q \in (0, 1)$; first, we will determine the left-hand side of Equation (16) and Equation (17) is obtained.

$$\begin{aligned} (1-q)\mathcal{N}(t, X, u) + q\mathcal{N}(t, X, v) =& p_1 B u_1 + p_2 B u_2 - c_1 u_1^2 - c_2 u_2^2 - c_3 u_3^2 - c_4 u^2 \\ & - q p_1 B u_1 - q p_2 B u_2 + q c_1 u_1^2 + q c_2 u_2^2 + q c_3 u_3^2 + q c_4 u^2 \\ & + q p_1 B v_1 + q p_2 B v_2 - q c_1 v_1^2 - q c_2 v_2^2 - q c_3 v_3^2 - q c_4 v^2, \end{aligned} \tag{17}$$

now, we develop the right-hand side of the inequality the Equation (16), Equation (18) is obtained.

$$\begin{aligned} \mathcal{N}(t, X, (1-q)u + qv) =& p_1 B u_1 - p_1 q B u_1 + p_1 q B v_1 + p_2 B u_2 - p_2 q B u_2 + p_2 q B v_2 \\ & - c_1[(1-q)u_1 + q v_1]^2 - c_2[(1-q)u_2 + q v_2]^2 \\ & - c_3[(1-q)u_3 + q v_3]^2 - c_4[(1-q)u_4 + q v_4]^2, \end{aligned} \tag{18}$$

developing each of the quadratic terms is obtained Equation (19).

$$\begin{aligned} c_1[(1-q)u_1+qv_1]^2 &= c_1u_1^2 - c_1qu_1^2 + c_1[(q^2-q)u_1^2 + 2q(1-q)u_1v_1 + q^2v_1^2] \\ c_2[(1-q)u_2+qv_2]^2 &= c_2u_2^2 - c_2qu_2^2 + c_2[(q^2-q)u_2^2 + 2q(1-q)u_2v_2 + q^2v_2^2] \\ c_3[(1-q)u_3+qv_3]^2 &= c_3u_3^2 - c_3qu_3^2 + c_3[(q^2-q)u_3^2 + 2q(1-q)u_3v_3 + q^2v_3^2] \\ c_4[(1-q)u_4+qv_4]^2 &= c_4u_4^2 - c_4qu_4^2 + c_4[(q^2-q)u_4^2 + 2q(1-q)u_4v_4 + q^2v_4^2], \end{aligned} \tag{19}$$

then, replacing Equation (19) in Equation (18), Equation (20) is obtained.

$$\begin{aligned} \mathcal{N}(t,X,(1-q)u+qv) =& p_1Bu_1 - p_1qBu_1 + p_1qBv_1 + p_2Bu_2 - p_2qBu_2 + p_2qBv_2 \\ & - [c_1u_1^2 - c_1qu_1^2 + c_1[(q^2-q)u_1^2 + 2q(1-q)u_1v_1 + q^2v_1^2]] \\ & - [c_2u_2^2 - c_2qu_2^2 + c_2[(q^2-q)u_2^2 + 2q(1-q)u_2v_2 + q^2v_2^2]] \\ & - [c_3u_3^2 - c_3qu_3^2 + c_3[(q^2-q)u_3^2 + 2q(1-q)u_3v_3 + q^2v_3^2]] \\ & - [c_4u_4^2 - c_4qu_4^2 + c_4[(q^2-q)u_4^2 + 2q(1-q)u_4v_4 + q^2v_4^2]], \end{aligned} \tag{20}$$

now, replacing Equations (17) and Equation (20) in Equation (16), we obtain the Equation (21).

$$(1-q)\mathcal{N}(t,X,u) + q\mathcal{N}(t,X,v) - \mathcal{N}(t,X,(1-q)u+qv) = -\sum_{i=1}^{4} c_i\left[\sqrt{q(1-q)}u_i - \sqrt{q(1-q)}v_i\right]^2 < 0, \tag{21}$$

as $e^{-\delta t} > 0$, we obtain the Equation (22).

$$(1-q)\mathcal{N}(t,X,u) + q\mathcal{N}(t,X,v) - \mathcal{N}(t,X,(1-q)u+qv) < 0, \tag{22}$$

then, the integrand of the objective functional $\mathcal{N}(t,X,u)$ is concave.
(v) Finally, as $e^{-\delta t} > 0$ is complied, in Equation (23) we get the required inequality.

$$\begin{aligned} \mathcal{N}(t,X,u) &\leq p_1FB + p_2TB - c_1F^2 - c_2T^2 - c_3R^2 - c_4S^2 \\ &\leq d_2 - d_1(|F|^2 + |T|^2 + |R|^2 + |S|^2). \end{aligned} \tag{23}$$

The Equation (23) is satisfied if $c_1 > \frac{p_1}{2}$ and $c_2 > \frac{p_2}{2}$; where d_2 depends on B, p_1 and p_2, while $d_1 = \min(\frac{(2c_1-p_1)}{2}, \frac{(2c_2-p_2)}{2}, c_3, c_4)$ and $d_3 = 2$. Thus, the assumptions of Theorem 1 are satisfied; we conclude that there exists an optimal control quadruple $J(F^*, T^*, R^*, S^*)$ such that $J(F,T,R,S)$ is maximized.

3. Conclusions

In this paper we present a bioeconomic optimal control model, which maximizes the net benefit of the timber volume; the mathematical model is related to three state variables and four control variables. We conclude that the optimal control problem is well formulated because the conditions of the Filippov Theorem are satisfied; there is thus an optimal control quadruple that maximizes the objective functional J.

Future expectations are to determine the necessary and sufficient conditions for the characterizations of the control variables utilizing the Pontryagin maximum principle and numerical simulations.

4. Acknowledgments

A. Altamirano would like to thank the Vicerrectoría de Investigación y Postgrado at Universidad Católica del Maule, Chile. This work is part of A. Altamirano Ph.D. thesis in the program of Doctorado en Modelamiento Matemático Aplicado.

References

[1] Cleophas F, *et al* 2022 Effect of soil physical properties on soil infiltration rates *Journal of Physics: Conference Series* **2314(1)** 012020
[2] Vásquez-Coronel J, *et al* 2022 Prediction of survival of pinus radiata seedlings subjected to physical-water restriction extreme using learning neural networks *Journal of Physics: Conference Series* **2153(1)** 012015
[3] Bull G Q, Bazett M, Schwab O, Nilsson S, White A and Maginnis S 2006 Industrial forest plantation subsidies: impacts and implications *Forest Policy and Economics* **9(1)** 13
[4] Cossalter C and Pye-Smith C 2003 *Fast-Wood Forestry. Myths and Realities* (Indonesia: Center for International Forestry Research)
[5] Reyes R and Nelson H 2014 A tale of two forests: why forests and forest conflicts are both growing in chile *International Forestry Review* **16(4)** 379
[6] Corriga G, Salimbeni D, Sanna S and Usai G 1988 An optimal control problem in forest management *Applied Mathematical Modelling* **12(3)** 328
[7] Gong Z, O'Hara K L, Li W and Gu L 2019 Optimal forest rotation periods: integrating timber production and carbon sequestration benefits in pinus tabulaeformis plantations on the loess plateau, pr china *Journal of Sustainable Forestry* **38(6)** 591
[8] Liu L *et al* 2022 Optimum rotation age of chinese fir (cunninghamia lanceolate) plantations under faustmann and maximum sustainable yield criteria *Journal of Tropical Forest Science* **34(3)** 305
[9] Menage D, Chetehouna K and Mell W 2012 Numerical simulations of fire spread in a pinus pinaster needles fuel bed *Journal of Physics: Conference Series* **395(1)** 012011
[10] Pasalodos-Tato M, Pukkala T, Castedo-Dorado F *et al.* 2009 Models for the optimal management of pinus radiata d. don in galicia (north-western spain) under risk of fire *Allgemeine Forst und Jagdzeitung* **180(11/12)** 238
[11] Banitalebi A, Aziz M I A, Aziz Z A and Nasir N 2016 Modelling and optimization for palm oil plantation management *AIP Conference Proceedings* **1750(1)** 030046
[12] Nasir N, Abd Aziz M I and Banitalebi A 2019 Carbon absorption control model of oil palm plantation *Sains Malaysiana* **48(4)** 921
[13] Altamirano-Fernández A, Rojas-Palma A and Espinoza-Meza S 2022 A mathematical model to study the dynamics of carbon capture in forest plantations *Journal of Physics: Conference Series* **2159(1)** 012001
[14] Moreno N, Moreno R and Molina J 2019 Optimal harvest cycle on nothofagus forests including carbon storage in southern america: an application to chilean subsidies in temperate forests *Land Use Policy* **81** 705
[15] Nyakundi F, Mulwa R and Kabubo-Mariara J 2018 Determination of optimal rotation period for management of lumbering forests in kenya *Journal of Sustainable Forestry* **37(7)** 645
[16] González-González J M, Vázquez-Méndez M E and Diéguez-Aranda U 2022 Multi-objective models for the forest harvest scheduling problem in a continuous-time framework *Forest Policy and Economics* **136** 102687
[17] Gaoue O G, Ngonghala C N, Jiang J and Lelu M 2016 Towards a mechanistic understanding of the synergistic effects of harvesting timber and non-timber forest products *Methods in Ecology and Evolution* **7(4)** 398
[18] Panja P and Jana D K 2020 A novel prey-predator quadratic harvesting model via optimal control theory and hopf bifurcation *Nepal Journal of Mathematical Sciences* **1** 1
[19] Alon U 2007 Chapman & hall/crc mathematical and computational biology series *An Introduction to Systems Biology: Design Principles of Biological Circuits* (Boca Raton: Chapman & Hall)
[20] Cesari L 1983 Lagrange and bolza problems of optimal control and other problems *Optimization—Theory and Applications. Problems with Ordinary Differential Equations* (New York: Springer) p 196
[21] Huong V T 2022 Solution existence theorems for finite horizon optimal economic growth problems *Optimization* **71(14)** 4243
[22] Fleming W and Rishel R 1975 Stochastic differential equations and markov diffusion processes *Deterministic and Stochastic Optimal Control* (New York: Springer) p 106
[23] Lukes D L 1982 *Differential Equations: Classical to Controlled* (New York: Academic Press)
[24] Khan M, Ali K, Bonyah E, Okosun K, Islam S and Khan A 2017 Mathematical modeling and stability analysis of pine wilt disease with optimal control *Scientific Reports* **7(1)** 3115
[25] Logan J D 2008 *An Introduction to Nonlinear Partial Differential Equations, 2nd Edition* (New Jersey: John Wiley & Sons)

Journal of Physics: Conference Series **2515** (2023) 012002 doi:10.1088/1742-6596/2515/1/012002

Consequences of insecticides on the dynamics of Anopheles mosquitoes

E Gómez-Hernández[1], E H Salazar-Jurado[1,2], A Altamirano-Fernández[1], and R Hernández-García[2,3]

[1] Facultad de Ciencias Básicas, Universidad Católica del Maule, Talca, Chile

[2] Laboratorio de Investigación Tecnológica en Reconocimiento de Patrones, Universidad Católica del Maule, Talca, Chile

[3] Centro de Investigación de Estudios Avanzados del Maule, Universidad Católica del Maule, Talca, Chile

E-mail: enith.gomez@alu.ucm.cl, rhernandez@ucm.cl

Abstract. Physics is a science that has developed valuable contributions to the understanding of different phenomena; in particular, the concept of a dynamic system originates in Newtonian mechanics; a concept that allows predictions of phenomena that are changing over time. The evolution rule is an implicit relation that can be given by a differential equation; our goal is to use dynamic systems in a problem of global importance, such as malaria. This is a potentially fatal disease caused by parasites transmitted to humans by biting infected female mosquitoes of the genus Anopheles; control is mainly due to combating the vectors. However, extensive use of insecticides has subjected mosquitoes to intense selection pressure, resulting in the development of physiological and behavioural resistance; therefore, we propose to study the dynamics of resistant and non-resistant infected mosquitoes. Our contribution is a system of ordinary differential equations in which non-resistant mosquitoes are assumed to have a logistic growth and exit upon developing resistance. In addition, we assume that resistant mosquitoes can have non-resistant offspring and only natural death; we performed a stability analysis of the model, allowing us to conclude that both types of mosquitoes always exist beyond a certain threshold. In addition, we performed a numerical study considering different levels of insecticide represented by the deaths it produces.

1. Introduction

Physics has contributed to understanding epidemic processes of infectious diseases by incorporating its tools in analyzing the dynamics of epidemics. The models used in the study of diseases generally have dynamics similar to those studied in physics, including oscillatory solutions, phase transitions, and pattern formation in time. This indicates that methodologies similar to those used in physics are helpful in this study. This interdisciplinarity allows us to be interested in studying malaria from the dynamical systems approach [1]. Malaria is considered one of the oldest diseases of humanity. About 3.2 billion people are exposed, making it a public health problem. In 2020 there were an estimated 241 million cases. In this regard, the World Health Organization (WHO) is constantly developing strategies to combat it; a plan is currently in place to eradicate malaria, mainly in Asia and South America, from 2016-2025 [2].

In South America, the most affected countries are Brazil, Peru and Venezuela, which report the highest number of annual cases, with an infected percentage of 77% of all cases in the region [3]. Several strategies have been developed for malaria control, among the most important of which is the control of mosquito vectors with insecticide applications, particularly with pyrethroids that are harmless to humans and other mammals but toxic to mosquitoes. In addition, they are cheap, long-lasting, and practical, as they kill and repel mosquitoes at low doses. However, insecticide resistance threatens to compromise this strategy [4]. Insecticide resistance has been identified among Anopheles mosquitoes, malaria vectors, in 64 countries with active malaria transmission, affecting all regions covered by the WHO [2, 5].

Given insecticide interventions important role in reducing the burden of malaria transmission, insecticide application requires knowledge of vector population dynamics, particularly their susceptibility status to the significant pyrethroids used for control programs [6, 7]. If no action is taken and insecticide resistance eventually results in widespread pyrethroid failure, much of the progress made in reducing malaria transmission will be lost [8]. For that reason, WHO has developed a global plan for resistance management in malaria vectors to ensure the effectiveness of insecticides as a current and future tool in preventing malaria morbidity and mortality [2]. An investigation of insecticide resistance can have as preliminary stage a dynamic system that allows us to determine critical parameters for decision-making in future experimental work. These systems can show how to correctly manage a pyritroids application threshold [9, 10].

The study of insecticide resistance using dynamic systems has provided results that contribute significantly to this problem. It has been found that if insecticide resistance levels continue to increase, epidemics may re-emerge. Furthermore, it has been concluded that in some cases, it is favorable to stop eliminating mosquitoes with insecticides before they become resistant because once they are resistant, transmission can increase rapidly [11]. The classical dynamic systems to study resistance consider the interactions between mosquitoes and humans, as in the case of the article [12], in which they describe the dynamics of the complete cycle of the parasite that causes malaria, incorporating the impact of the intensity of resistance to insecticides. However, it is possible to find papers in the literature that only consider the mosquito population. This is the case of [13], in which a mathematical model predicts that insecticide-resistant mosquitoes will become dominant over time but that it is essential to occur over a long time.

This article aims to study the dynamics of infected mosquitoes resistant and non-resistant to insecticides. We found a threshold of population growth that allows us to qualitatively analyze the system. Furthermore, we employ sensitivity analysis to determine the influence of parameters such as growth rate, natural death rate, and the fraction of female mosquitoes. Most importantly, we evaluate the effect of parameters associated with insecticide resistance, such as the rate at which non-resistant mosquitoes acquire resistance and insecticide-induced mortality. In this sense, the contribution to physics is related to a complex dynamical system that presents organizational principles similar to any other dynamical system that can serve as a tool in other fields.

2. The model

We propose a mathematical model based on ordinary differential equations to study the dynamics of resistant and non-resistant infected mosquitoes; both populations with standard incidence and taking into account the following classes: non-resistant infected mosquito (x) and resistant infected mosquito(y), the total population M is given by: $M = x + y$. Figure 1 shows the flowchart of the above dynamics and takes into account the following assumptions.

(i) Non-resistant infected mosquitoes x have a logistic growth and output flow due to the development of resistance to a rate α and mortality rate μ_1 (insecticide application).

(ii) In non-resistant mosquitoes, we consider the fraction of female mosquitoes f and a growth rate γ.

(iii) Resistant infected mosquitoes y have an output flow by a natural death rate ϵ.

Journal of Physics: Conference Series **2515** (2023) 012002 doi:10.1088/1742-6596/2515/1/012002

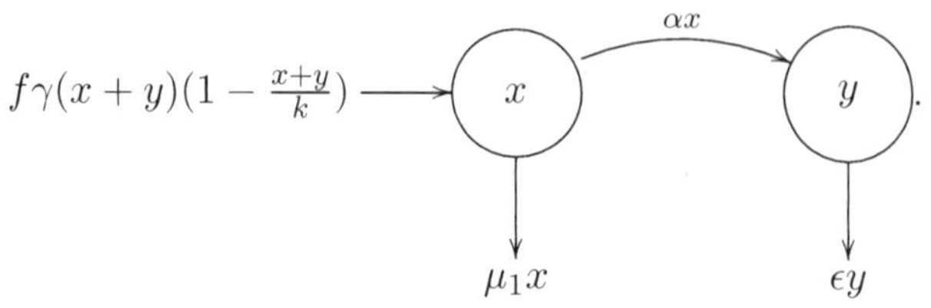

Figure 1. Flux diagram of the dynamics.

Then the dynamics can be written as a system of nonlinear ordinary differential equations, as shown in Equation (1).

$$\begin{cases} \dfrac{dx}{dt} = f\gamma(x+y)\left(1-\dfrac{x+y}{k}\right) - (\alpha+\mu_1)x, \\ \dfrac{dy}{dt} = \alpha x - \epsilon y, \end{cases} \tag{1}$$

where f, γ, k, α, μ_1, ϵ are positive, and the initial conditions are $x(0) = x_0$, $y(0) = y_0$. The region of epidemiological sense is $\Omega \in \mathbb{R}^2_+$, it which is positively invariant for the system presented in Equation (1), and it is given by Equation (2).

$$\Omega = \{(x,y) \in \mathbb{R}^2 : x \geq 0, y \geq 0, x+y \leq k\}, \tag{2}$$

The Equation (2) assures that the populations are always positive and do not grow indefinitely in time. The following propositions show results about the local and asymptotic stability.

Proof. Always there exists that equilibrium point without resistant and non-resistant mosquitoes $E_0 = (0,0)$ and an equilibrium point with resistant and non-resistant mosquitoes $E_1 = (\hat{x}, \hat{y})$ which have an epidemiological sense region Ω when the threshold is $\Phi \leq 1$ and $\Phi > 1$ respectively.

Proof. Hence equaling to zero the right-hand side of Equation (1), we obtain Equation (3) and solve for x.

$$x\left[f\gamma\left(1+\frac{\alpha}{\epsilon}\right)\left(1-\frac{\left(1+\frac{\alpha}{\epsilon}\right)x}{k}\right) - (\alpha+\mu_1)\right] = 0, \tag{3}$$

the solutions of Equation (3) are expressed in Equation (4).

$$x = 0 \quad \vee \quad x = \frac{k(\epsilon+\mu_1)}{f\gamma\left(1+\frac{\alpha}{\epsilon}\right)^2}(\Phi-1), \tag{4}$$

where, Φ is defined in Equation (5) and represents the growth threshold of the mosquitoes population.

$$\Phi = \frac{f\gamma(\epsilon+\alpha)}{\epsilon(\alpha+\mu_1)}. \tag{5}$$

The growth threshold can be rewritten by Equation (6) for better understanding.

$$\Phi = \frac{f\gamma}{(\alpha+\mu_1)} + \frac{f\gamma\alpha}{\epsilon(\alpha+\mu_1)}. \tag{6}$$

The first term is the female mosquitoes fraction f with a growth rate γ multiplied by the life period of non-resistant infected mosquitoes $1/(\alpha+\mu_1)$, the right side on the expresion are the non-resistant infected mosquitoes which develop a resistance rate α. By replacing Equation (4) in $y=\frac{\alpha}{\epsilon}x$ are obtained the equilibrium points $E_0=(0,0)$ and $E_1=(\hat{x},\hat{y})$, where $\hat{x}$ and $\hat{y}$ are defined in Equation (7).

$$\hat{x}=\frac{k\epsilon(\Phi-1)}{(\epsilon+\alpha)\Phi} \quad \wedge \quad \hat{y}=\frac{k\alpha(\Phi-1)}{(\epsilon+\alpha)\Phi}. \tag{7}$$

Proof. Where $\Phi<1$ the equilibrium point $E_0=(0,0)$ is local and stable asymptotically in Ω.

Proof. The Jacobian matrix J at point $E_0=(0,0)$ is given in Equation (8).

$$J(E_0)=\begin{pmatrix} f\gamma-(\alpha+\mu_1) & f\gamma \\ \alpha & -\epsilon \end{pmatrix}, \tag{8}$$

where, the determinant (δ) and the trace (τ) from $J(E_0)$ are given in Equation (9) and Equation (10), respectively.

$$\delta = -\epsilon\,(f\gamma-(\alpha+\mu_1))-\alpha\gamma f, \tag{9}$$
$$\tau = f\gamma-(\alpha+\mu_1+\epsilon). \tag{10}$$

From Equation (9) is obtained $-\epsilon f\gamma+\epsilon(\alpha+\mu_1)-\alpha\gamma f>0$ which implies the Equation (11) and Equation (12).

$$-\frac{\epsilon f\gamma}{\epsilon(\alpha+\mu_1)}+\frac{\epsilon(\alpha+\mu_1)}{\epsilon(\alpha+\mu_1)}-\frac{\alpha\gamma f}{\epsilon(\alpha+\mu_1)} \Leftrightarrow -\frac{f\gamma(\epsilon+\alpha)}{\epsilon(\alpha+\mu_1)}+1 > 0, \tag{11}$$
$$\Phi < 1. \tag{12}$$

And from Equation (10) it is obtained the inequalities of Equation (13) and Equation (14).

$$f\gamma<(\alpha+\mu_1)+\epsilon \Leftrightarrow \frac{f\gamma}{\alpha+\mu_1}<1+\frac{\epsilon}{\alpha+\mu_1} \Leftrightarrow \frac{f\gamma(\epsilon+\alpha)}{\epsilon(\alpha+\mu_1)} < \frac{\epsilon+\alpha}{\epsilon}+\frac{\epsilon+\alpha}{\alpha+\mu_1} \tag{13}$$
$$\Phi < 1+\frac{\alpha}{\epsilon}+\frac{\epsilon+\alpha}{\alpha+\mu_1}. \tag{14}$$

Thus $\Phi<1$.

Proof. When $\Phi>1$ and $k<2(\hat{x}+\hat{y})$ the equilibrium point $E_1=(\hat{x},\hat{y})$ is local and asymptotically stable in Ω.

Proof. The Jacobian matrix is obtained at the generic point $E_1=(\hat{x},\hat{y})$ and it is given in Equation (15).

$$J(E_1)=\begin{pmatrix} A-(\alpha+\mu_1) & A \\ \alpha & -\epsilon \end{pmatrix}, \tag{15}$$

where, $A=f\gamma\,(1-(2(\hat{x}+\hat{y}))/k)$ and the determinant (δ) and the trace (τ) from the matrix $J(E_1)$ are given in Equation (16) and Equation (17), respectively.

$$\delta = -\epsilon(A - (\alpha + \mu_1)) - \alpha A, \tag{16}$$
$$\tau = A - (\alpha + \mu_1) - \epsilon. \tag{17}$$

If $A < 0$ then $k < 2(\hat{x} + \hat{y})$ in which case the Equation (18) and Equation (19) are obtained.

$$\delta = -\epsilon A + \epsilon(\alpha + \mu_1) - \alpha A > 0, \tag{18}$$
$$\tau = A - (\alpha + \mu_1) - \epsilon < 0. \tag{19}$$

Thus, $E_1 = (\hat{x}, \hat{y})$ is local and asymptotically stable for $\Phi > 1$ and $k < 2(\hat{x} + \hat{y})$.

3. Sensitivity analysis

The sensitivity analysis of the threshold Φ and the coexistence equilibrium E_1 in terms of the Equation (1) parameters are obtained with information from Table 1. The sensitivity analysis determines how the parameters have influence each other, through small variations on the parameter in which the sensitivity index has greater magnitude, namely: if a parameter ν depends on ω, its sensitivity index is given by $I_\omega^\nu = \frac{\omega}{\nu}\frac{\partial \nu}{\partial \omega}$.

Table 1. Simulation parameters.

Parameter	f	γ	ϵ	α	μ_1
Value	0.600	0.530	0.083	0.300	0.300 - 0.600 - 0.900

3.1. Sensitivity index of Φ

It is important to reduce the mosquito growth threshold Φ, thus it is necessary a sensitivity analysis respecting to f, γ, ϵ, α parameters and μ_1. Then, $I_f^\Phi = 1$, $I_\gamma^\Phi = 1$, $I_\epsilon^\Phi = \frac{\epsilon}{\epsilon+\alpha} - 1$, $I_\alpha^\Phi = \frac{\alpha}{\epsilon+\alpha} - \frac{\alpha}{\alpha+\mu_1}$ and $I_{\mu_1}^\Phi = -\frac{\mu_1}{\alpha+\mu_1}$. In the Table 2 the numerical values shown for each index.

Table 2. Sensitivity index of Φ.

Parameter	f	γ	ϵ	α	μ_1
Sensitivity index	1	1	-0.78	0.1	-0.50

The most sensitivity parameters for Φ are f and γ, so, the female mosquitoes fraction and the growth rate are directly proportional to the growth threshold and Φ decreases when the natural death rate of the resistant infected mosquitoes ϵ increases, see Figure 2.

3.2. Coexistence equilibrium

The equilibrium point E_1 in terms of Equation (1) parameters is given for Equation (20).

$$\hat{x} = \frac{k\epsilon(\Phi - 1)}{(\epsilon + \alpha)\Phi} \quad \wedge \quad \hat{y} = \frac{k\alpha(\Phi - 1)}{(\epsilon + \alpha)\Phi}. \tag{20}$$

If the growth threshold Φ is greater than 1 while time passes then the mosquitoes population approaches to E_1, taking into account this condition, the sensitivity index of $\hat{x}$, $\hat{y}$ respecting the parameters are given in Equation (21) and Equation (22). In the Table 3 the numerical values shown for each index.

$$I_{\epsilon}^{\hat{x}} = \frac{\alpha\phi}{\phi-1} - \frac{2\alpha}{(\epsilon+\alpha)(\phi-1)}, \; I_{\alpha}^{\hat{x}} = \frac{-\alpha(\phi(\alpha+\mu_1)+\epsilon-2\mu_1-\alpha)}{(\epsilon+\alpha)(\alpha+\mu_1)(\phi-1)}, \; I_{\gamma}^{\hat{x}} = I_{f}^{\hat{x}} = I_{\gamma}^{\hat{y}} = I_{f}^{\hat{y}} = \frac{1}{\phi-1}. \quad (21)$$

$$I_{\mu_1}^{\hat{x}} = I_{\mu_1}^{\hat{y}} = \frac{-\mu_1}{(\alpha+\mu_1)(\phi-1)}, \quad I_{\epsilon}^{\hat{y}} = \frac{-\epsilon(\phi+1)+\alpha}{(\epsilon+\alpha)(\phi-1)}, \quad I_{\alpha}^{\hat{y}} = \frac{\phi\epsilon(\alpha+\mu_1)+\mu_1(\alpha-\epsilon)-2\epsilon\alpha}{(\alpha+\mu_1)(\epsilon+\alpha)(\phi-1)}. \quad (22)$$

Table 3. Sensitivity index of $\hat{x}$, $\hat{y}$.

Parameter	ϵ	α	f	μ_1
Sensitivity index $\hat{x}$	0.24	-0.58	0.69	-0.34
Sensitivity index $\hat{y}$	-0.75	0.41	0.69	-0.34

The most sensitivity parameters in the non-resistant infected mosquitoes $\hat{x}$ are the female mosquitoes fraction f and the growth rate γ on the other hand in the resistant infected mosquitoes population $\hat{y}$ the most sensitive parameter is the natural death rate of resistant mosquitoes ϵ, as shown in the Figure 2.

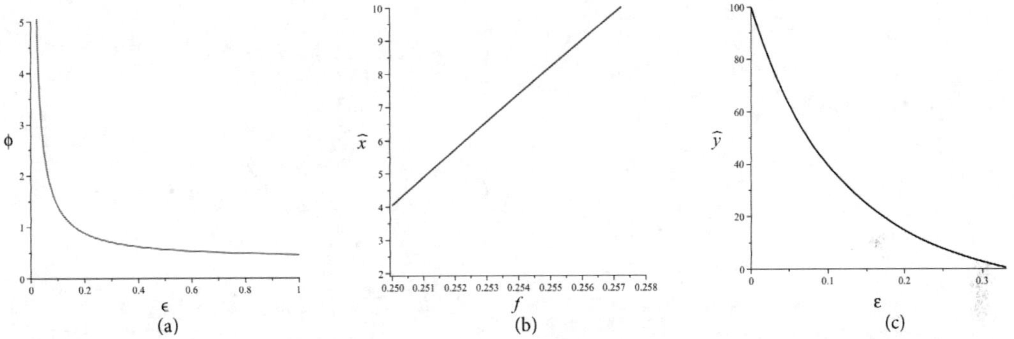

Figure 2. Sensitivity index; (a) sensitivity index of the growth threshold ϕ as a function of the natural death ϵ, (b) sensitivity index of non-resistant infected mosquitoes $\hat{x}$ at equilibrium as a function of the fraction of female mosquitoes f, (c) sensitivity index of resistant infected mosquitoes $\hat{y}$ at equilibrium as a function as a function of the natural death ϵ.

4. Simulations

The simulations were carried out using the parameters of Table 1, with a charge capacity $k = 1000$ and initial conditions $x(0) = 100$, $y(0) = 10$. We have used diferent values of the insecticide induce mortality rate obtaining that the behavior for each μ_1 es very similar; however, when the mortality for insecticide application increases, the number of resistant and non-resistant mosquitoes decreases, see Figure 3.

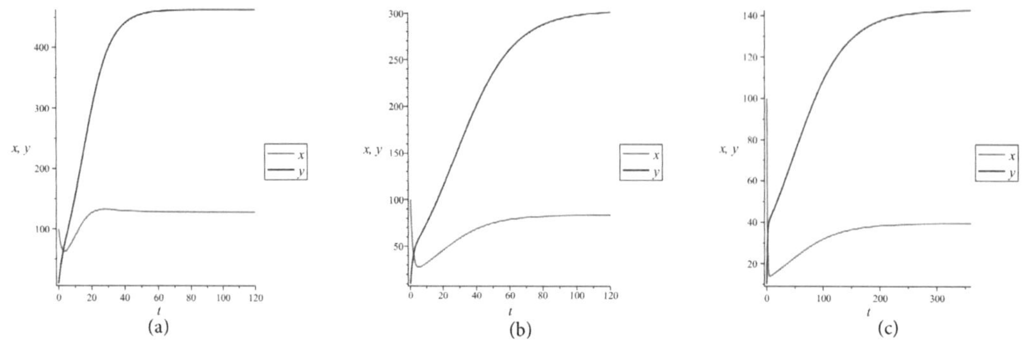

Figure 3. Behavior of resistant y and non-resistant x infected mosquitoes at time t, changing the insecticide induced mortality rate μ_1. (a) $\mu_1 = 0.3$. (b) $\mu_1 = 0.6$. (c) $\mu_1 = 0.9$.

In Figure 4 is depicted the first phase diagram, which was realized taking information from Table 1, excluding the natural death rate value; for this case $\epsilon = 0.83$ is used and then $\Phi < 1$, showing that the behavior of the species stabilizes at the equilibrium point E_0. The second diagram was done as well taken information from Table 1, obtaining $\Phi > 1$ assures the stability at the equilibrium point E_1.

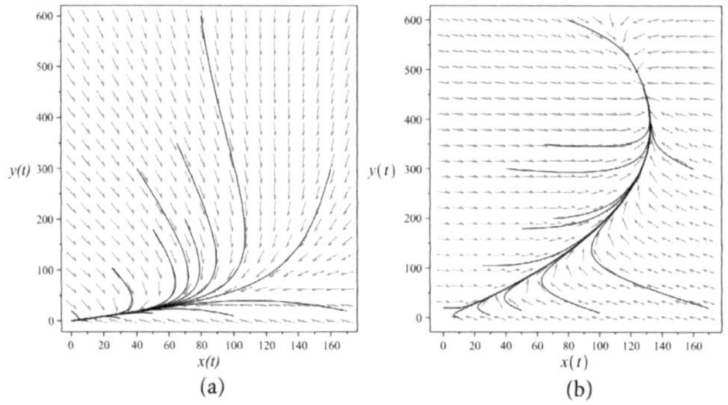

Figure 4. Phase diagrams; (a) $\Phi < 1$, (b) $\Phi > 1$.

5. Conclusions

The qualitative analysis shows two equilibrium solutions, local and asymptotically stable, depending on the threshold Φ; also, it has been found that if the number of insecticide-induced deaths increases, the growth of both populations decreases. In this scenario, the non-resistant mosquitoes population decreases quickly. However, after this one acquires immunity, the resistant population rises. Future work is to consider a model with a population of malaria-infected and uninfected humans and determine the influence of resistance on the system's behavior.

References

[1] Wang Z, Moreno Y, Boccaletti S and Perc M 2017 Vaccination and epidemics in networked populations—an introduction *Chaos, Solitons & Fractals* **103** 177–183
[2] Organization W H 2019 *World Malaria Report* (Geneva: World Health Organization)
[3] Organization P A H 2017 *Health in the Americas+* (Washington: Regional Outlook and Country Profiles)
[4] Müller P, Donnelly M J and Ranson H 2007 Transcription profiling of a recently colonised pyrethroid resistant anopheles gambiae strain from ghana *BMC Genomics* **8(1)** 1–12

[5] Fonseca-González I, Cárdenas R, Gómez W, Santacoloma L, Brochero H, Ocampo C, Salazar M, MCALLIS TER J, Brogdon W and Quiñones M 2010 Dosis diagnósticas para vigilar la resistencia a insecticidas de los vectores de malaria en colombia *Revista Colombiana de Entomología* **36(1)** 54–61

[6] Machani M G, Ochomo E, Amimo F, Mukabana W R, Githeko A K, Yan G and Afrane Y A 2022 Behavioral responses of pyrethroid resistant and susceptible anopheles gambiae mosquitoes to insecticide treated bed net *PLoS ONE* **17(4)** e0266420

[7] Morgan J C, Irving H, Okedi L M, Steven A and Wondji C S 2010 Pyrethroid resistance in an anopheles funestus population from uganda *PLoS ONE* **5(7)** e11872

[8] Matowo N S, Martin J, Kulkarni M A, Mosha J F, Lukole E, Isaya G, Shirima B, Kaaya R, Moyes C, Hancock P A *et al.* 2021 An increasing role of pyrethroid-resistant anopheles funestus in malaria transmission in the lake zone, tanzania *Scientific Reports* **11(1)** 1–13

[9] Saddler A and Koella J C 2015 Modelling the impact of declining insecticide resistance with mosquito age on malaria transmission *Malaria World Journal* **6(13)** 1–7

[10] Barbosa S and Hastings I M 2012 The importance of modelling the spread of insecticide resistance in a heterogeneous environment: the example of adding synergists to bed nets *Malaria Journal* **11(1)** 1–12

[11] Griffin J T, Bhatt S, Sinka M E, Gething P W, Lynch M, Patouillard E, Shutes E, Newman R D, Alonso P, Cibulskis R E *et al.* 2016 Potential for reduction of burden and local elimination of malaria by reducing plasmodium falciparum malaria transmission: a mathematical modelling study *The Lancet Infectious Diseases* **16(4)** 465–472

[12] Nwankwo A 2021 Quantifying the impact of insecticide resistance in the transmission dynamics of malaria *Chaos, Solitons & Fractals* **142** 110481

[13] Gourley S A, Liu R and Wu J 2011 Slowing the evolution of insecticide resistance in mosquitoes: a mathematical model *Proceedings of the Royal Society A: Mathematical, Physical and Engineering Sciences* **467(2132)** 2127–2148

Journal of Physics: Conference Series **2515** (2023) 012003 doi:10.1088/1742-6596/2515/1/012003

Estimation of the optimal number of neurons in extreme learning machine using simulated annealing and the golden section

E Gelvez-Almeida[1,2], M Mora[1], Y Huérfano-Maldonado[1], E Salazar-Jurado[1], N Martínez-Jeraldo[3], R Lozada-Yavina[3,4], Y Baldera-Moreno[3], and L Tobar[1]

[1] Laboratorio de Investigaciones Tecnológicas en Reconocimiento de Patrones, Universidad Católica del Maule, Talca, Chile
[2] Facultad de Ciencias Básicas y Biomédicas, Universidad Simón Bolívar, San José de Cúcuta, Colombia
[3] Facultad de Ciencias Básicas, Universidad Católica del Maule, Talca, Chile
[4] Facultad de Ciencias e Ingeniería, Universidad Tecnológica del Perú, Lima, Perú

E-mail: mmora@ucm.cl, elkin.gelvez@unisimon.edu.co

Abstract. Extreme learning machine is a neural network algorithm widely accepted in the scientific community due to the simplicity of the model and its good results in classification and regression problems; digital image processing, medical diagnosis, and signal recognition are some applications in the field of physics addressed with these neural networks. The algorithm must be executed with an adequate number of neurons in the hidden layer to obtain good results. Identifying the appropriate number of neurons in the hidden layer is an open problem in the extreme learning machine field. The search process has a high computational cost if carried out sequentially, given the complexity of the calculations as the number of neurons increases. In this work, we use the search of the golden section and simulated annealing as heuristic methods to calculate the appropriate number of neurons in the hidden layer of an Extreme Learning Machine; for the experiments, three real databases were used for the classification problem and a synthetic database for the regression problem. The results show that the search for the appropriate number of neurons is accelerated up to 4.5× times with simulated annealing and up to 95.7× times with the golden section search compared to a sequential method in the highest-dimensional database.

1. Introduction

One of the significant challenges facing science in classification and regression problems is the high computational cost of working with large-dimensional databases; more efficient models have been developed to mitigate these problems without losing accuracy. Among the single hidden layer artificial neural network (ANN) models are the extreme learning machines (ELM) [1], whose fundamental characteristic is their good performance concerning computation time required for its training against the position with other networks; the main procedure in ELM training is to randomize the hidden layer weights and calculate the output layer weights using the Moore-Penrose generalized inverse.

The ELM is a widely used model for classification problems, regression, and clustering [1]. When this model performs the training process, a database percentage is taken to train and another to test the accuracy. When this process is done, in order to obtain good precision, the appropriate number of neurons in the hidden layer must be identified. For the above, optimization methods are required; for example, sequentially. it would take a lot of computation time and is inefficient. The accuracy of the ELM can improve when there is an adequate number of neurons in the hidden layer; however, its search is slow if we do sequential tests one by one. Optimization methods can significantly speed up the search process without compromising accuracy. It happens as fewer iterations are performed compared to sequential search.

Various mechanisms have been proposed to find the appropriate number of neurons in the hidden layer, such as optimally pruned [2, 3], algorithms based on Bayesian optimization [4, 5], particle swarm optimization (PSO) algorithms [6–8], genetic algorithms [9–11], heuristic methods from the laws of electromagnetism [12], among others. However, the general trend is to use complex optimization algorithms, which allow finding excellent values of hyperparameters but with a high execution time, moving away from the philosophy and advantages of ELM networks.

In this work, we have used two simple optimization methods to search for the appropriate number of neurons in the hidden layer of the ELM; the methods are simulated annealing and the golden section, compared with the commonly used sequential search. Simulated annealing is a heuristic method inspired by metal annealing and is based on the Boltzmann probability distribution [13]; in the case of the golden section, it is an iterative process that seeks to maximize or minimize an objective function using the golden ratio [14]. Finally, in the sequential search, all the central processing unit (CPU) cores have been used to make the most of the Hardware architecture. Four public databases were considered for the experiments: one for binary classification tasks, two for multiclass classification tasks, and one for regression tasks.

2. Background

ELM is a learning algorithm for single hidden layer feedforward networks (SLFN) [1]; these neural networks are used in classification, regression, and clustering tasks. It is noteworthy that the training time in the ELM is significantly less compared to other popular algorithms such as back-propagation (BP) and support vector machines (SVM) [1].

2.1. Moore–Penrose generalized inverse

The solution of an overdetermined system of linear equations $\boldsymbol{Ax} = \boldsymbol{y}$, where $\boldsymbol{A} \in \mathbb{R}^{m\times n}$ with $m, n \in \mathbb{N}$ and $m > n$, can be determined using the generalized Moore-Penrose inverse $\boldsymbol{A}^\dagger \in \mathbb{R}^{n\times m}$ [15]; this inverse is unique and satisfies the four conditions of Equation (1), called Penrose conditions.

$$\boldsymbol{AA}^\dagger\boldsymbol{A} = \boldsymbol{A}, \quad \boldsymbol{A}^\dagger\boldsymbol{AA}^\dagger = \boldsymbol{A}^\dagger, \quad \boldsymbol{A}^\dagger\boldsymbol{A} = (\boldsymbol{A}^\dagger\boldsymbol{A})^T, \quad \text{and} \quad \boldsymbol{AA}^\dagger = (\boldsymbol{AA}^\dagger)^T. \tag{1}$$

2.2. Extreme learning machine

Let $\aleph$ be an arbitrary training set $\{(\boldsymbol{x}_i, \boldsymbol{t}_i)|\boldsymbol{x}_i \in \mathbb{R}^n, \boldsymbol{t}_i \in \mathbb{R}^m\}$ with $i = 1, \ldots, N$, an activation function $g : \mathbb{R}^n \to \mathbb{R}^m$, and L number of neurons in the hidden layer with $L < N$, the SLFN training algorithm is given by Equation (2) [1].

$$\sum_{i=1}^{L} \boldsymbol{\beta}_i g(\boldsymbol{w}_i \boldsymbol{x}_j + b_i) = \boldsymbol{t}_j, \qquad j = 1, \ldots, N. \tag{2}$$

In Equation (2), $\boldsymbol{w}_i$ and b_i are, respectively, the i-th weights and biases of the hidden layer, $\boldsymbol{\beta}_i$ the i-th weight of the output layer, and $\boldsymbol{w}_i \cdot \boldsymbol{x}_j$ represents the inner product of $\boldsymbol{w}_i$ and $\boldsymbol{x}_j$ [16].

Equation (2) can be written in the matrix form $\boldsymbol{H}\boldsymbol{\beta} = \boldsymbol{T}$, where $\boldsymbol{H}$ is called the output matrix of the hidden layer of the neural network [1].

2.3. Optimization methods

The methods used in this article aim to quickly and efficiently search for the optimal number of neurons in the hidden layer of an ELM; the algorithms used are described below.

2.3.1. Simulated annealing. It is a heuristic method inspired by the annealing of metals; its algorithm is based on the logic of gradual cooling to thermal equilibrium [17]. This behavior is studied based on the Boltzmann probability distribution, given by Equation (3); this distribution establishes that the probability $P[\delta_E]$ of energy jump δ_E in a given solid decreases exponentially as a function of the quotient between the energy jump δ_E and the product of the temperature t and Boltzmann's constant K.

$$P[\delta_E] = \exp^{-\frac{\delta_E}{tK}}. \tag{3}$$

According to Equation (3), the probability of an energy jump δ_E decreases for high energy values and low temperatures t. Kirkpatrick *et al.* [13] introduced the idea that this behavior can be used to solve optimization problems in optimization processes; the objective function is equivalent to the energy jump of metal annealing. Therefore, the admissible solutions move in a wide margin for high temperatures while the acceptance process decreases for low temperatures, thus favoring the escape of local minima.

2.3.2. Golden section search. It is an optimization method that seeks to maximize or minimize a function and is applied when the objective function is unimodal [14]; the method calculates the points x_1 and x_2 of an interval $[l, u]$ using Equation (4).

$$\begin{aligned} x_1 &= l + (1 - d) * (u - l), \\ x_2 &= l + d * (u - l). \end{aligned} \tag{4}$$

The parameter d of Equation (4) is derived from the golden ratio and is estimated using Equation (5).

$$d = \frac{1 + \sqrt{5}}{2} - 1. \tag{5}$$

Once the values of x_1 and x_2 are found, the function is evaluated on l, x_1, x_2, and u; if the minimum of the function is sought, it must be taken into account that if $f(x_1) < f(x_2)$, then the minimum is in the interval $[l, x_2]$. Otherwise, the minimum is in the interval $[x_1, u]$; this process continues with the new bounds until the stopping criterion [14] is met. The absolute value gives the stop criterion between the extremes, defined according to the precision obtained. The number of iterations required depends on the stopping criterion.

3. Materials and methods

For the experiments, a computer with an Intel(R) Core(TM) i7-8550U 1.80GHz–1.99GHz central processing unit (CPU) (×4 cores) and 12Gb of RAM were used. The codes were made in MATLAB R2017a; below is a description of this article's data sets and methods.

3.1. Description of the databases
Four public databases are used to evaluate the algorithms described in the previous section; diabetes database [18] is used for binary classification problems, and the samples are characterized by eight attributes, which determine if a person has diabetes. The sat-image database is used for multiclass classification problems, and its samples are obtained with the Landsat multispectral scanner (orbiting satellites for high-resolution observation of the Earth's surface) [19]; shuttle database is used for multiclass classification problems just like the sat-image database [20].

In this case, the samples are described by nine attributes containing information related to a spacecraft's landing conditions; lastly, unlike the previous databases, sinc database [16] is an artificial database used for regression problems. The training samples are perturbations of the cardinal sine function, while the testing samples correspond to values of the function without perturbation; Table 1 summarizes the characteristics of the databases used.

Table 1. Public databases used in the optimal computation of neurons for ELM.

Database	Application	Classes	Attributes	Training	Testing
Diabetes [18]	Classification	2	8	576	129
Satimage [19]	Classification	6	36	4435	2000
Shuttle [20]	Classification	7	9	43500	14500
Sinc [16]	Regression	-	2	5000	5000

3.2. Optimization process
In the first instance, a sequential search was carried out, increasing the number of neurons in the hidden layer up to 1,200 neurons; the same number of neurons in the hidden layer has been selected to make the experiments identical in all the databases. Once the results are obtained, the ELM algorithm is executed using the optimization methods described in Section 2.3; these results are compared with the sequential search to identify which optimization method requires fewer iterations and execution time. The following section analyzes the results obtained.

4. Results
The results obtained in the sequential search are first considered to carry out the comparison; Figure 1 shows the error curve as a function of the number of neurons for each database described in Section 3.1. It can be seen in the sat-image and shuttle database that the error decreases as the number of neurons in the hidden layer increases until converging on a minimum error. However, with the diabetes database, the error is greater when the number of neurons in the hidden layer increases and stabilizes when the number of samples equals; for its part, in the sinc database, a regression problem, the error is stabilized with a few neurons in the hidden layer.

Once the results of the sequential search are known, the optimization methods are executed. Table 2 shows the number of iterations performed by each method on the different databases; in addition, the minimum error obtained and the number of neurons necessary for that result can be observed. Finally, the time in minutes the optimization methods take to obtain these results is presented; the times associated with the sequential search were obtained using the MATLAB parfor command to take full advantage of the four processor cores.

The execution time of the sequential search and optimization methods is proportional to the number of samples and neurons in the hidden layer; the shuttle database has the largest number of training samples. Therefore the sequential search and optimization methods are executed at higher times, especially when compared to the diabetes database, whose difference between samples is significant.

Journal of Physics: Conference Series **2515** (2023) 012003 doi:10.1088/1742-6596/2515/1/012003

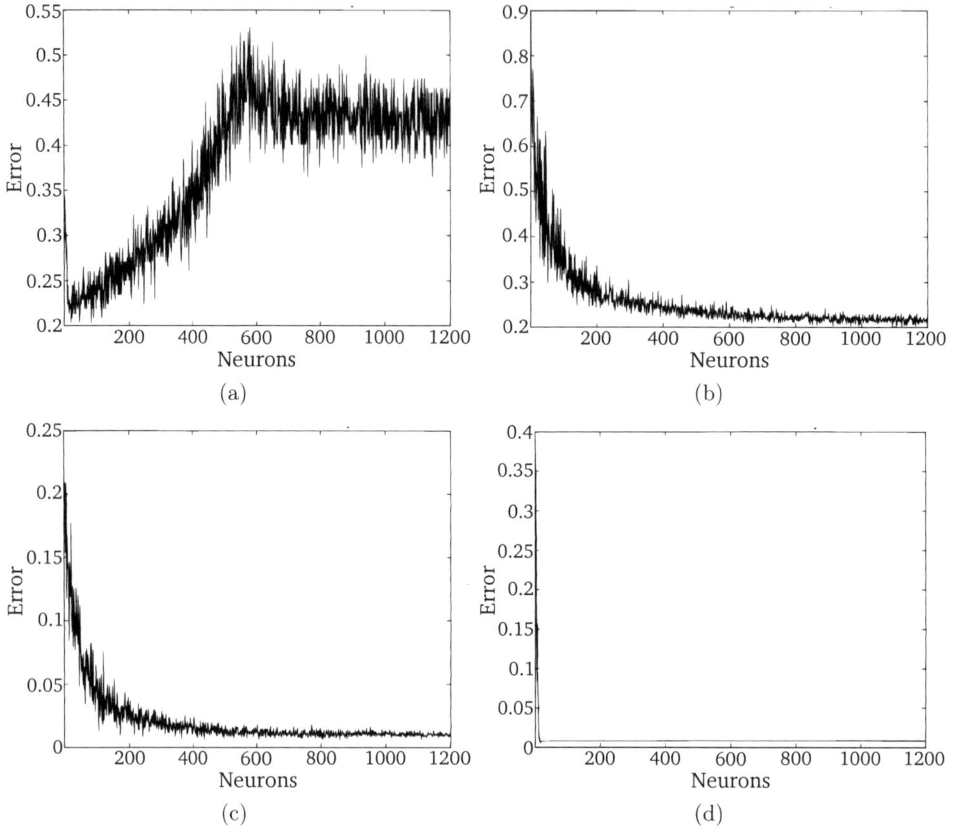

Figure 1. ELM error as the number of neurons in the hidden layer increases; (a) results with diabetes database, (b) results with the sat-image database, (c) results with shuttle database, (d) results with sinc database.

Table 2. Results obtained with the different databases and optimization methods.

Database	Method	Iterations	Error	Neurons	Time (min)
Diabetes	Sequential (parfor)	1200	0.2031	21, 47, 83	1.15
	Simulated annealing	523	0.2031	44	0.33
	Golden section	34	0.2396	22	0.03
Satimage	Sequential (parfor)	1200	0.2005	1094	11.17
	Simulated annealing	1049	0.2140	432	2.66
	Golden section	34	0.2050	1042	1.36
Shuttle	Sequential (parfor)	1200	0.0061	858	104.38
	Simulated annealing	576	0.0054	397	24.77
	Golden section	31	0.0093	192	1.09
Sinc	Sequential (parfor)	1200	0.0055	15	10.24
	Simulated annealing	506	0.0071	17	2.26
	Golden section	34	0.0085	823	0.87

As can be seen in the results, execution times are significantly reduced when using optimization methods compared to sequential search, mainly due to the number of iterations each method performs. It is worth highlighting the results obtained with the golden section method due to the significant reduction in the number of iterations; for its part, the minimum error obtained by each optimization method is very close to that obtained with the sequential search. The similarity of the error is because the optimization methods stop their execution at a minimum error using convergence criteria, in the case of the golden section and simulated annealing.

Regarding each database, the results show that in the diabetes database, an acceleration of 3.48× and 38.33× is achieved with the simulated annealing and the golden section, respectively, compared to the sequential search. In the sat-image database, simulated annealing speeds up the search to 4.20× times, while the golden section speeds up 8.22× times. The biggest difference is obtained in the shuttle database, speeding up the search 4.21× times with the simulated annealing and 95.76× times with the golden section. Results in the shuttle database are much higher than in the other databases due to the number of training samples. Finally, in the sinc database, a speedup of 4.53× and 11.77× is achieved with the simulated annealing and the golden section, respectively. In general terms, the search for the golden section finds, in less time, an optimal number of neurons in the hidden layer without affecting the precision.

5. Conclusions

We have used simulated annealing and the golden section to determine the optimal number of neurons in the hidden layer of an extreme learning machine and, additionally, reduce computation time; in the experiments, it can be observed that the methods used constitute a valid alternative for the automatic detection of the optimal number of neurons required for the efficient and effective training of the extreme learning machine. The comparison has been made with a sequential search that maximizes the computer's hardware, specifically, the multicore central processing unit; the effectiveness of simulated annealing is that it performs a deep, forward search, ensuring an optimal number of neurons in the hidden layer. However, when performing many iterations, its computational cost increases; for its part, the golden section performs fewer iterations in determining an optimal number of neurons, which reduces the time cost compared to simulated annealing and sequential search.

Acknowledgments

This work was funded by the Agencia Nacional de Investigación y Desarrollo (ANID) / scholarship program / 2020-21201000 / 2021-21211760 / 2022-21220765. This paper is also one of the scientific results of the research project Fondecyt regular 2020 N° 1200810 "Very large fingerprint classification based on a fast and distributed extreme learning machine", Agencia Nacional de Investigación y Desarrollo (ANID), Chile. E G-A also thanks for the licenses for doctoral studies, the "Fondo para el desarrollo docente y profesional" of the Universidad Simón Bolívar, Colombia. N M-J, Y B-M, and R L-Y thanks to the "doctoral scholarship", awarded by the Vicerectoría de Investigación y Postgrado of the Universidad Católica del Maule, Chile.

References

[1] Huang G B, Zhu Q Y and Siew C K 2004 Extreme learning machine: a new learning scheme of feedforward neural networks *Proceedings of the 2004 IEEE International Joint Conference on Neural Networks (IEEE Cat. No. 04CH37541)* vol 2 (Budapest: IEEE) p 985

[2] Miche Y, Sorjamaa A, Bas P, Simula O, Jutten C and Lendasse A 2009 OP-ELM: optimally pruned extreme learning machine *IEEE Transactions on Neural Networks* **21(1)** 158

[3] Miche Y, Van Heeswijk M, Bas P, Simula O and Lendasse A 2011 TROP-ELM: a double-regularized elm using lars and tikhonov regularization *Neurocomputing* **74(16)** 2413

[4] Soria-Olivas E, Gomez-Sanchis J, Martin J D, Vila-Frances J, Martinez M, Magdalena J R and Serrano A J 2011 BELM: bayesian extreme learning machine *IEEE Transactions on Neural Networks* **22(3)** 505
[5] Thornton C, Hutter F, Hoos H H and Leyton-Brown K 2013 Auto-WEKA: combined selection and hyperparameter optimization of classification algorithms *Proceedings of the 19th ACM SIGKDD International Conference on Knowledge Discovery and Data Mining* ed Ghani R, Senator T E and Bradley P (Chicago: Association for Computing Machinery) p 847
[6] Ahila R, Sadasivam V and Manimala K 2015 An integrated PSO for parameter determination and feature selection of elm and its application in classification of power system disturbances *Applied Soft Computing* **32** 23
[7] Ali M H, Fadlizolkipi M, Firdaus A and Khidzir N Z 2018 A hybrid particle swarm optimization-extreme learning machine approach for intrusion detection system *Proceedings of the 2018 IEEE Student Conference on Research and Development (SCOReD)* (Selangor: IEEE) p 1
[8] Bao X, Li Y, Li J, Shi R and Ding X 2021 Prediction of train arrival delay using hybrid ELM-PSO approach *Journal of Advanced Transportation* **2021** 7763125
[9] Cao Z, Xia J, Zhang M, Jin J, Deng L, Wang X and Qu J 2015 Optimization of gear blank preforms based on a new R-GPLVM model utilizing GA-ELM *Knowledge-Based Systems* **83** 66
[10] Ahmad W, Ayub N, Ali T, Irfan M, Awais M, Shiraz M and Glowacz A 2020 Towards short term electricity load forecasting using improved support vector machine and extreme learning machine *Energies* **13(11)** 2907
[11] Martinho A D, Ribeiro C, Gorodetskaya Y, Fonseca T L and Goliatt L 2020 Extreme learning machine with evolutionary parameter tuning applied to forecast the daily natural flow at cahora bassa dam, mozambique *Bioinspired Methods and Their Applications* ed Filipič B, Minisci E and Vasile M (Cham: Springer) p 255
[12] Zhang M, Wu Q, Xu Z *et al.* 2019 Tuning extreme learning machine by an improved electromagnetism-like mechanism algorithm for classification problem *Mathematical Biosciences and Engineering* **16(5)** 4692
[13] Kirkpatrick S, Gelatt C D and Vecchi M P 1983 Optimization by simulated annealing *Science* **220(4598)** 671
[14] Koupaei J A, Hosseini S M M and Ghaini F M 2016 A new optimization algorithm based on chaotic maps and golden section search method *Engineering Applications of Artificial Intelligence* **50** 201
[15] Lyche T 2020 *Numerical Linear Algebra and Matrix Factorizations* vol 22 (Oslo: Springer)
[16] Huang G B, Zhu Q Y and Siew C K 2006 Extreme learning machine: theory and applications *Neurocomputing* **70(1-3)** 489
[17] Haznedar B, Arslan M T and Kalinli A 2021 Optimizing ANFIS using simulated annealing algorithm for classification of microarray gene expression cancer data *Medical & Biological Engineering & Computing* **59(3)** 497
[18] Kubat M 1999 Neural networks: a comprehensive foundation *The Knowledge Engineering Review* vol 13 ed Simon H (New York: Cambridge University Press) p 409
[19] Peltonen J, Klami A and Kaski S 2004 Improved learning of riemannian metrics for exploratory analysis *Neural Networks* **17(8-9)** 1087
[20] Hsu C W and Lin C J 2002 A comparison of methods for multiclass support vector machines *IEEE Transactions on Neural Networks* **13(2)** 415

Mathematical model of polylactic acid biodegradation under controlled composting conditions

Y Baldera-Moreno[1], A Rojas-Palma[1], R Andler[2], and L Cuesta-Herrera[1]

[1] Doctorado en Modelamiento Matemático Aplicado, Universidad Católica del Maule, Talca, Chile

[2] Escuela de Ingeniería en Biotecnología, Universidad Católica del Maule, Talca, Chile

E-mail: yvan.baldera@alu.ucm.cl

Abstract. The environment suffers constant damage due to the excessive manufacture of non-degradable materials. Since petroleum-derived plastic polymers degrade slowly, it is necessary to promote the use of biodegradable plastics, such as polylactic acid. Polylactic acid is produced from natural and renewable resources, it has suitable physical and chemical properties, and biodegrades under appropriate temperature and humidity conditions, which are achieved in the composting process. Currently, there are models that are based on the measurement of carbon dioxide, and mass loss, among others. However, there are no models that describe the biodegradation of polylactic acid considering the effect of the influencing factors of the composting process. The objective of this work is to model the dynamics of polylactic acid biodegradation under controlled composting conditions, considering the main influencing factors, such as temperature, moisture content, and oxygen. Using mathematical modeling from ordinary differential equations as a methodology, simulations were carried out based on the degradation of dry matter from different substrates. The results aim to predict the dynamics of polylactic acid biodegradation, through a model that integrates the influencing factors of the composting process.

1. Introduction

Plastics are synthetic polymers with a complex macromolecular structure that can be molded by heat or pressure with carbon (C) being the main component [1]. These polymers are large groupings of small molecules (monomers) united in a repeating formation through a chemical process called polymerization [2].

Plastics are classified according to their manufacturing origin: petroleum-derived (synthetic) and bio-based (natural) polymers. Petroleum-derived plastics have a high production worldwide, of which the most important are: polyethylene (PE), polyethylene terephthalate (PET), polyvinyl chloride (PVC), polypropylene (PP), polystyrene (PS) and polyurethane (PU). All these plastics can remain for hundreds of years in the environment, because they have a high prevalence of carbon-carbon (C-C) bonds, especially in PE, PP, PVC, and PS, whereas, PU has a urethane bond and PET is an ester bond, both of which can be hydrolyzed [3]. Bio-based plastics are made from biological and renewable resources, such as polylactic acid (PLA) and polyhydroxyalkanoates (PHA), starch- and cellulose-based plastics. This group of plastics degrades at higher rates since the corresponding ester and glycosidic bonds are easily catalyzed by microorganisms, which makes polymer fragmentation (into oligomers or monomers) fast [4].

According to ISO 14855 [5], plastic is biodegradable if it reaches 90% biodegradation in a maximum time of 180 days under controlled composting conditions. The latter is a biological process that occurs under aerobic conditions (presence of oxygen) which depends on influencing factors such as temperature, moisture content, oxygen, aeration, carbon/nitrogen ratio (C/N), hydrogen potential (pH), among others [6]. PLA is a biodegradable bio-based plastic produced from natural and renewable resources such as corn, wheat, cassava, and sugar cane, commonly used due to its availability, biodegradability, and good mechanical attributes, and whose hydrolytic degradation by-products can be completely digested by microorganisms. [7]. For these reasons, PLA has been selected as the plastic polymer to be studied in this work.

Mathematical modeling of the biodegradation kinetics of different plastics has progressed in recent years and is mainly used to monitor their biodegradability [8]. In the literature it is possible to find different models to evaluate the degradability of plastic materials, among which the following stand out: a) The measurement of carbon dioxide (CO_2) released by the microorganisms [9, 10] after metabolizing and mineralizing the micromolecules (monomers) obtained during depolymerization; b) The analysis of the mass loss of plastic samples at the beginning and at the end of the biodegradation process [11, 12]; c) The use of the Arrhenius equation to evaluate the effect of temperature variation on the biodegradation process [13, 14]; d) The use of the Michaelis-Menten equation to understand the enzymatic kinetics of depolymerization reactions [15, 16].

Despite the previously mentioned advances, from our knowledge, there are no studies of mathematical models that approach the biodegradation of plastics by integrating the different factors of influence the composting process. Therefore, our research objective is focused on modeling the biodegradation dynamics of PLA under controlled composting conditions. For this, the main influencing factors, such as temperature, moisture content, and oxygen availability were considered and the modeling was based on the use of ordinary differential equations.

2. Methodology

The main equations used for the formulation of the mathematical model proposed in this work are shown below.

2.1. Hydrolytic degradation of polylactic acid

Piemonte and Gironi [17] proposed the kinetic model of PLA hydrolytic degradation, Equation (1), taking into account two reaction phases (slow and fast), which is given by.

$$\frac{dm_{PLA}(t)}{dt} = -\frac{k_I m_{PLA}(t)}{1+e^{(t-t^*)}} - \frac{k_{II} m_{PLA}(t)(m_{PLA,0} - m_{PLA}(t))^{0.5}}{1+e^{(t^*-t)}}, \tag{1}$$

where, $m_{PLA}(t)$ is the mass concentration of PLA at time t, $m_{PLA,0}$ is the initial concentration of PLA, t^* is the switching time between the slow and fast reaction phase, k_I and k_{II} are the slow and fast reaction kinetic constants, respectively. In that work, the switching time, $t^* = a \cdot T + b$, depended on the temperature (T).

In Figure 1(a), the kinetics of PLA degradation is observed; the first phase is performed at a slow reaction rate $k_I = 0.000622$ (day^{-1}), the PLA polymeric chains are broken into low molecular weight fragments (oligomers or monomers) and culminate when achieving the switching time $t^* = 40$ days. In the second phase, PLA degradation occurs at a fast reaction rate $k_{II} = 0.112$ ($kg^{-0.5}\, day^{-1}$) is higher than the previous one because microorganisms start to metabolize the small particles of the degraded PLA.

2.2. Multicomponent model for the degradation of composting dry mass

Keener *et al.*, [18] formulated a multi-component model of compost dry mass, considering various substrates. In that study, analytical expressions were derived, which show the interdependence between biological and physical influencing factors. The following substrates were used to estimate the parameters of the equations: municipal solid waste, garden waste, poultry manure, and urban sewage sludge. The active phase of composting is the time when the substrate is available for microbial decomposition; during this time, the dry mass degradation rate, Equation (2), written as a first-order reaction, is given by.

$$\frac{dm_c(t)}{dt} = -\, k[m_c(t) - m_e], \tag{2}$$

where, m_c is the dry mass of composting, k is the degradation rate and m_e is the equilibrium. If k is constant, the solution of Equation (2), is given by.

$$mR = \frac{m_c(t) - m_e}{m_0 - m_e} = e^{-kt}, \tag{3}$$

where mR is the compost mass ratio. Equation (3) is a dimensionless number, which is a useful way to describe the degree of progress of the composting process. For a multi-component system (where several substrates are considered), the model would be given by.

$$\frac{dm_c(t)}{dt} = \sum_j \frac{dm_{jc}(t)}{dt} = \sum_j -k_j[m_{jc}(t) - m_{je}], \tag{4}$$

whose solution is $mR = u_1 e^{-k_1 t} + u_2 e^{-k_2 t} + \ldots$, from where $u_1 + u_2 + \ldots = 1.0$ and $u_j = \dfrac{m_{j0} - m_{je}}{m_0 - m_e}$.

In Figure 1(b), we can observe the negative exponential behavior that has the degradation curve of the compost dry mass (m_c), which is made up of three substrates (m_j, with $j = 1, 2, 3$). Considering that the degradation rates k_j (day^{-1}) are constant, *i.e.*, $k_1 = 0.0338, k_2 = 0.016, k_3 = 0.0176$; the equilibria of each substrate (m_{je}) are given by $m_{1e} = 0, m_{2e} = 0.399, m_{3e} = 0.02$ (see Equation (4)).

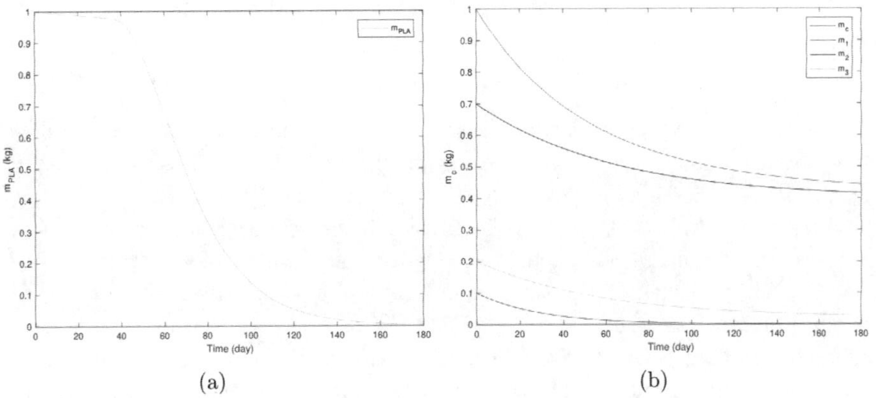

Figure 1. (a) Kinetics of hydrolytic degradation of PLA. (b) Composting model with three components.

Journal of Physics: Conference Series **2515** (2023) 012004 doi:10.1088/1742-6596/2515/1/012004

2.3. Mathematical model

The mathematical model that describes the biodegradation of PLA under controlled composting conditions is presented below, involving its main influencing factors (temperature (T), moisture content (m_{cw}) and oxygen concentration (C_{O_2})), which is given by Equation (5) to Equation (11).

$$\begin{aligned}\frac{dm_c(t)}{dt} &= -k_f[m_f(t)-m_{e,f}]-k_s[m_s(t)-m_{e,s}]-\frac{k_I m_{PLA}(t)}{1+e^{(t-t^*)}}\\ &- \frac{k_{II} m_{PLA}(t)(m_{PLA,0}-m_{PLA}(t))^{0.5}}{1+e^{(t^*-t)}},\end{aligned} \tag{5}$$

$$\frac{dm_f(t)}{dt} = -k_f[m_f(t)-m_{e,f}], \tag{6}$$

$$\frac{dm_s(t)}{dt} = -k_s[m_s(t)-m_{e,s}], \tag{7}$$

$$\frac{dm_{PLA}(t)}{dt} = -\frac{k_I m_{PLA}(t)}{1+e^{(t-t^*)}} - \frac{k_{II} m_{PLA}(t)(m_{PLA,0}-m_{PLA}(t))^{0.5}}{1+e^{(t^*-t)}}, \tag{8}$$

$$\frac{dT(t)}{dt} = \frac{1}{m_c c_p}\left(h_c\frac{dm_c}{dt} + \dot{m}_{in}HAI(\cdot) - \dot{m}_{out}HAO(\cdot) - UA(T-T_{amb})\right), \tag{9}$$

$$\frac{dm_{cw}(t)}{dt} = b_{cw}\frac{dm_c}{dt} + w_{as,in}(\cdot)\dot{m}_{in} - w_{as,out}(\cdot)\dot{m}_{out}, \tag{10}$$

$$\frac{dC_{O_2}(t)}{dt} = \frac{1}{\varepsilon V}\left(b_{O_2}\frac{dm_c}{dt} + A\nu_{in}(\cdot)C_{O_2,in} - A\nu_{out}(\cdot)C_{O_2,out}\right), \tag{11}$$

where k_j is the rate of degradation of each substrate and is given by Equation (12).

$$k_j = k_{j,max}\cdot X_T \cdot X_{m_{cw}} \cdot X_{O_2}, \qquad j = f, s, I, II, \tag{12}$$

whose correction factors are described by Equation (13).

$$X_T = e^{-0.5\left(\left[\frac{T-57.33}{15.82}\right]^2\right)}, \quad X_{m_{cw}} = e^{-0.5\left(\left[\frac{m_{cw}-44.51}{18.73}\right]^2\right)}, \quad X_{O_2} = 0.614\cdot\arctan(0.4\cdot C_{O_2}). \tag{13}$$

Equation (5) to Equation (8) describe the degradation kinetics of the dry compost mass (m_c) and its three components, *i.e.*, fast degrading substrate mass (m_f), slow degrading substrate mass (m_s) and PLA mass (m_{PLA}) as the third substrate. Note that, the biodegradation of m_{PLA} is divided into two reaction phases, firstly, because the microorganisms adapt to the polymer to fragment it, and upon reaching the switching time (t^*) they start to metabolize and mineralize the small particles of the degraded PLA. Moreover, these phases are performed at a slow degradation rate (k_I) and a fast degradation rate (k_{II}). In this model, the influencing factors of the composting process, Equation (9) to Equation (11), are involved to study their effect on each degradation rate k_j, Equation (12). The correction factors, Equation (13), of these degradation rates have been taken from secondary sources [19–21].

3. Results and discussions

The results obtained based on simulations corresponding to the Equation (5) to Equation (11), which consists of seven ordinary differential equations are presented. In addition, these simulations are discussed to test the importance of the physical properties of PLA and the effect that the influencing factors of the composting process have on its biodegradation. The parameter values used within the model simulations have been taken from secondary sources [17–27] (see Table 1).

Table 1. Parameters of PLA biodegradation dynamics under controlled composting conditions.

Symbol	Value	Unit	Description
b_{cw}	0.51	kg water kg air^{-1}	Water production coefficient
b_{O_2}	1.177	kg O_2 kg compost^{-1}	Utilization rate of O_2 of the compost mix
$C_{O_2}(0)$	21	kg O_2 m^{-3}	Initial oxygen concentration
h_c	18090.486	kJ kg compost^{-1}	Heat of combustion of the compost mixture
k_f	0.0337	day^{-1}	Rapid degradation rate of m_f
k_s	0.0135	day^{-1}	Slow degradation rate of m_s
k_I	0.0112	day^{-1}	Slow degradation rate of m_{PLA}
k_{II}	0.1755	day^{-1}	Rapid degradation rate of m_{PLA}
$m_{cw}(0)$	45	%w.b.	Initial moisture content
t^*	40	day	Switching time from slow reaction to fast reaction
$T(0)$	25.37	°C	Initial temperature
T_{amb}	23.31	°C	Ambient temperature
UA	4.5468	kJ d^{-1} °C^{-1}	Overall heat transfer coefficient per reactor area
ε	0.5	-	Compost porosity

In Figure 2(a), the degradation behavior of the dry composting matter (m_c) is described. This is composed of three substrates: paper mill sludge (fast degradation), broiler litter (slow degradation) and PLA (medium degradation). Unlike Figure 1(b), where the degradation rates were considered constant, the biodegradation curves of the substrates m_f, m_s and m_{PLA} were affected due to the effect of the correction factors, Equation (13), on their non-constant degradation rates (k_j), Equation (12). The dry matter degradation curve (m_c) of compost behaves similarly to the mass of PLA (m_{PLA}). This is due to the biodegradation process that PLA has to go through, which consists of three main phases. The first is known as the lag phase (0-40 days), where the microbial population adapts to the carbon substrate of the polymer and fragments it into smaller sections (oligomers or monomers). Second, the polymer begins the biodegradation phase (40-150 days), where the adapted microbial population begins to use the carbon substrate for its cellular metabolism processes. Third, the process culminates with the plateau phase (150-180 days), which is reached when the composting is stabilized and the PLA biodegradation is 90% in a maximum time of 180 days (according to norm ISO 14855 [5]).

Temperature (T), as one of the influencing factors, has three phases in the aerobic decomposition process. First, the initial mesophilic phase (T <40 °C) which takes place until the switching time of 40 days is reached. Second, the thermophilic phase (T >40 °C) occurs between approximately 40 – 70 days. Third, the process culminates with the final mesophilic phase, where the initial temperature of the compost is reached again (see Figure 2(b)). Whereas, the evolution of the moisture content curve (m_{cw}) was slightly affected by the changes of the temperature phases (see Figure 2(c)). For the correct development of composting it is necessary to ensure the presence of oxygen, since the microorganisms involved are aerobic. The oxygen concentration (C_{O_2}) begins to decrease slightly, until reaching the switching time where the initial mesophilic phase is terminated. Then, it decreases at a faster rate until the thermophilic phase ends and finally, when the composting process stabilizes, the oxygen concentration returns to the initial concentrations (see Figure 2(d)).

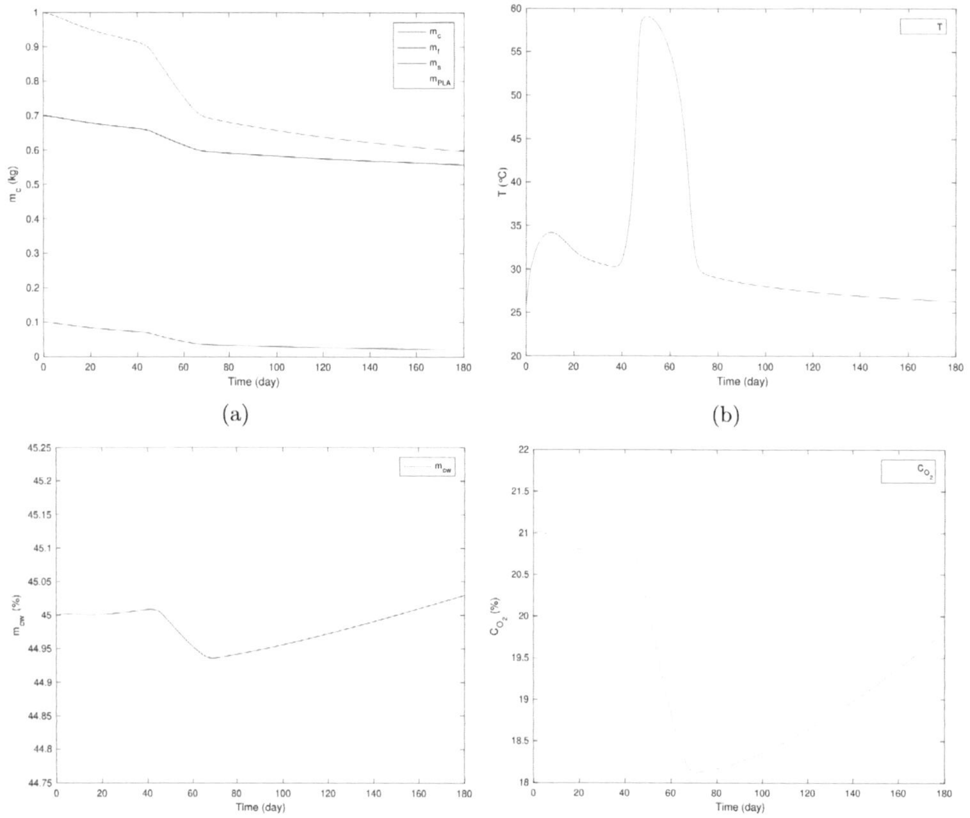

(a) (b) (c) (d)

Figure 2. Dynamics of PLA biodegradation under controlled composting conditions; (a) degradation of composting masses, (b) temperature, (c) moisture content, and (d) oxygen concentration.

4. Conclusions

In this paper, a system of ordinary differential equations to formulate a mathematical model that described the biodegradation of polylactic acid was used, involving the main influencing factors of the composting process (temperature, moisture content and oxygen concentration). These factors have an effect on the degradation rates of the three substrates considered: paper mill sludge (fast degradation), broiler litter (slow degradation) and polylactic acid (medium degradation).

The simulations performed show that the polylactic acid biodegradation in the composting process is 90%, in a maximum time of 180 days, confirming its biodegradability according to norm ISO 14855 and the secondary sources consulted.

This study shows that the polylactic acid biodegradation process is affected by different influencing factors, which are interrelated. Understanding the effect of these factors on the composting process could lead to a significant increase in polymer degradation rates. In addition, it helps to promote the use of biodegradable plastics, such as polylactic acid, and reduce dependence on petroleum-derived plastics.

Acknowledgments

Yvan Baldera Moreno and Ledyz Cuesta Herrera thank the "Doctoral Scholarship", awarded by the Vicerectoría de Investigación y Postgrado, Universidad Católica del Maule, Talca, Chile. Also, all the authors thank research project Fondecyt, grant number 11190220, from ANID, Chile.

References

[1] Laskar N and Kumar U 2019 *Environmental Technology & Innovation* **14** 100352
[2] Edmondson S and Gilbert M 2017 *Brydson's Plastics Materials* (United Kingdom: Elsevier) p 19
[3] Dhanraj N D, Hatha A A and Jisha M S 2022 *Archives of Microbiology* **204(5)** 1
[4] Valero-Valdivieso M F, Ortegon Y and Uscategui Y 2013 *Dyna* **80(181)** 171
[5] Funabashi M, Ninomiya F and Kunioka M 2009 *International Journal of Molecular Sciences* **10(8)** 3635
[6] Reyes-Torres M, Oviedo-Ocaña E, Dominguez I, Komilis D and Sánchez A 2018 *Waste Management* **77** 486
[7] Ahmed T, Shahid M, Azeem F, Rasul I, Shah A A, Noman M, Hameed A, Manzoor N, Manzoor I and Muhammad S 2018 *Environmental Science and Pollution Research* **25(8)** 7287
[8] Baldera-Moreno Y, Pino V, Farres A, Banerjee A, Gordillo F and Andler R 2022 *Polymers* **14(3)** 375
[9] Leejarkpai T, Suwanmanee U, Rudeekit Y and Mungcharoen T 2011 *Waste Management* **31(6)** 1153
[10] Gómez E F and Michel Jr F C 2013 *Polymer Degradation and Stability* **98(12)** 2583
[11] Ruggero F, Carretti E, Gori R, Lotti T and Lubello C 2020 *Chemosphere* **246** 125770
[12] Ruggero F, Onderwater R C, Carretti E, Roosa S, Benali S, Raquez J M, Gori R, Lubello C and Wattiez R 2021 *Journal of Polymers and the Environment* **29(9)** 3015
[13] 2020 *ACS Sustainable Chemistry & Engineering* **8(9)** 3494
[14] Perejón A, Sánchez-Jiménez P E, García-Garrido C and Pérez-Maqueda L A 2021 *Polymer Degradation and Stability* **188** 109590
[15] Farzi A, Dehnad A and Fotouhi A F 2019 *Biocatalysis & Agricultural Biotechnology* **17** 25
[16] Rossetti I, Conte F and Ramis G 2021 *Eng* **2(1)** 54
[17] Piemonte V and Gironi F 2013 *Journal of Polymers and the Environment* **21(2)** 313
[18] Keener H M, Hansen R C and Marugg C 1992 *Optimizing the Efficiency of the Composting Process* (Ohio: The Ohio State University)
[19] Ekinci K 2001 *Theoretical and Experimental Studies on the Effects of Aeration Strategies on the Composting Process* (Ohio: The Ohio State University)
[20] Ekinci K, Keener H, Michael F and Elwell D 2001 *Proceedings of ASAE Annual International Meeting* (California: ASAE) p 1
[21] Keener H, Elwell D, Michel F *et al.* 2005 *Transactions of the ASAE* **48(3)** 1203
[22] Ekinci K, Keener H M and Elwell D 2002 *Compost Science & Utilization* **10(1)** 16
[23] Ekinci K, Keener H, Michel F and Elwell D 2004 *Compost Science & Utilization* **12(4)** 356
[24] Keener H M, Ekinci K and Michel F C 2005 *Compost Science & Utilization* **13(4)** 288
[25] Mustafić N, Petrić I and Karić E 2017 *Journal of Engineering & Processing Management* **9(1)** 62
[26] Soyöz C, Ekinci K and Kilic Ş 2021 *Waste and Biomass Valorization* **12(7)** 3645
[27] Ekinci K, Keener H M and Akbolat D 2006 *Bioresource Technology* **97(7)** 922

Study of barley yellow dwarf virus: dynamics of aphid-plant interaction

E Gómez-Hernández[1], L Cuesta-Herrera[1], A Altamirano-Fernández[1], E Salazar-Jurado[1,2], and J Vásquez-Coronel[1,2]

[1] Doctorado en Modelamiento Matemático Aplicado, Universidad Católica del Maule, Talca, Chile

[2] Laboratorio de Investigación Tecnológica en Reconocimiento de Patrones, Universidad Católica del Maule, Talca, Chile

E-mail: enith.gomez@alu.ucm.cl, eamandagh@gmail.com

Abstract. Physics continues to be an inexhaustible source of problems that have inspired the theoretical development of mathematics, particularly in the development of the theory of dynamical systems that have their origin in the physics of the 15th century, with the beginning of differential calculus by Newton and Leibniz; differential equations are an appropriate tool for research in different areas of knowledge. Our objective is to show an example of how we can use dynamical systems in a problem framed in forestry and agricultural systems, ecology, and epidemiology. This work is inspired by one of the essential primary crops in the energy intake of the human diet, wheat; as it turns out, the productivity of this food can be affected by several biotic stresses, including viral diseases and their associated vectors. Among the viral infections is the barley yellow dwarf virus, which causes 11% to 33% yield losses in wheat fields; the virus moves through the agricultural landscape via different aphid species (vectors). The main problem is that aphids prefer colonizing plants with the virus, causing rapid growth of infected plants and substantially decreasing crop productivity; therefore, we propose to study the dynamics of aphid-plant interaction, in which the impact of aphid preference for infected plants is determined, and we obtained as a result that a key parameter is the quantity and quality of food consumed by aphids.

1. Introduction

Wheat is the most important staple crop in temperate zones and is in increasing demand in urbanizing and industrializing countries. Globally, it provides 20% of the energy intake of the human diet [1,2]. However, the productivity of this food can be affected by several biotic stresses, including viral diseases and their associated vectors [3]. Among viral infections, barley yellow dwarf virus (BYDV) is prominent, which causes 11% to 33% yield losses in wheat fields [4]. The virus moves across the agricultural landscape via different aphid species (vectors). The most recognized aphid species are Metopolophium dirhodum, Rhopalosiphum padi, Sitobion avenae, and Schizaphis graminum [5,6]. The virus restricts photosynthate transport and causes phloem degeneration within the plant, resulting in a reduced number of tillers, grains per ear, and grain weight, in addition to yellowing of leaves and dwarfing [5,7].

The BYDV is not transferred from infected aphids to their progeny, and aphids only acquire the virus by feeding on infected plants. Therefore, the smaller the number of infected plants, the greater the possibility of controlling the virus [5]. The problem is that aphids tend to be attracted by the yellow color, i.e., they prefer colonizing plants with the virus, causing rapid growth of infected plants and a substantial decrease in crop productivity [3,8]. This phenomenon has been studied experimentally, for example, by observing the food preference of two aphid species using chemical signals [9]. These studies are essential in the research process. However, due to the economic cost of experimental work, dynamical systems can be used as a first research resource. It is a resource that builds on existing knowledge about the phenomenon and, in this particular case, would allow us to know the impact of the tendency of aphids to feed on plants with the virus.

Some authors have already addressed different issues surrounding BYDV with dynamical systems [10–14]. In [10], they use an epidemiological model of virus spread to study the number of aphids and the population dynamics of three viral serotypes in the plant. The authors use the classical susceptible-infectious system used in humans. They consider the dynamics of susceptible plants and three infectious compartments depending on the serotypes. It is noteworthy that this is a work where the interaction with aphids was not modeled, and instead, they estimated a fixed parameter by sampling. In [11], they studied a particular case of complete unilateral cross-protection between two viruses; they considered two types of viruses in the plant, a protective virus and a virulent virus, under the assumption that the protective virus completely excludes the virulent virus, but not vice versa. They then determined the dynamics of four state variables: healthy plants, plants infected with the protective virus, plants infected with the virulent virus, and plants infected with both types of virus. The authors' objective was to investigate whether the artificial introduction of a protective virus is effective for disease control in crops.

In the previously mentioned works, the interaction of hosts and vectors was not included, unlike the papers [12, 13], in which aphids are state variables. In [12], they studied the competition between viruses in a complex plant-pathogen system. For this, they propose a system that considers plants and aphids. Plants are divided into four compartments: susceptible plants, dormant plants, and two compartments of infected plants to model two types of viruses. The aphids are divided into subpopulations: susceptible aphids, and two classes of infected aphids to differentiate between strains. With this study, the authors conclude that competition within the plant is more significant than competition within the vectors if you want to implement a disease control mechanism. In [13], they use a host plant-vector interaction model to show how the presence of natural enemies of aphids can increase disease development, even if vector density decreases. By studying their dynamical system, the authors suggest that biological control of aphids may increase disease incidence attributable to increased vector movement in response to the presence of other anthropods.

From a general perspective, we can say that each dynamical system is constructed according to a particular study objective. Therefore, in this paper, we formulate a system of ordinary differential equations to analyze the effect of aphid preference for infected plants. Our dynamic system has four state variables; host plants and aphids, both are divided into susceptible and infected following the classical epidemiology model. We assume that aphids have a logistic growth, in which their carrying capacity is variable and depends on plant density; in contrast, plant density growth has a logistic growth with a fixed carrying capacity. The interaction between susceptible aphids (infected aphids) and infected plants (susceptible plants) leads to an input flow on the state variable of infected aphids (infected plants) by means of a modification of the classical function used for disease transmission in which we include a parameter of relative preference of aphids to feed on infected plants.

2. The model

We build a model that tracks population dynamics of the susceptible aphids $X_S(t)$, infected aphids $X_I(t)$, susceptible host-plant $Y_S(t)$ and infected host-plant $Y_I(t)$ at time t; we make the following assumptions.

In the absence of aphids on the plants, the density of susceptible plants Y_S grows according to a logistic fashion with carrying capacity K ($K \in \mathbb{R}_+$), with a constant intrinsic birth rate r ($r \in \mathbb{R}_+$) and under the assumption that infected plants have susceptible offspring as shown in Equation (1).

$$\frac{dY_S}{dt} = r_2(Y_S + Y_I)\left(1 - \frac{Y_S + Y_I}{K}\right). \tag{1}$$

The BYDV transmission in the plant occurs when infected aphids feed on susceptible plants. Then we assume that there is an outflow from the Y_S compartment by virus transmission at a rate β_2 ($\beta_2 \in \mathbb{R}_+$). In addition, we formulate the transmission function according to [15], an article in which the authors assume that the density of the susceptible plants is calculated as $Y_S/(Y_S + \alpha Y_I)$, where α ($\alpha \in \mathbb{R}_+$) is the relative preference of aphids to feed on infected plants. Under these assumptions, the rate of change of Y_S is given by Equation (2).

$$\frac{dY_S}{dt} = r_2(Y_S + Y_I)\left(1 - \frac{Y_S + Y_I}{K}\right) - \beta_2 \frac{Y_S}{Y_S + \alpha Y_I} X_I. \tag{2}$$

The outflow of Equation (2) is precisely the inflow from the compartment of the infected plants Y_I as shown in Equation (3); it is noteworthy that we do not consider death by disease because the BYDV only decreases the productivity of the crop.

$$\frac{dY_I}{dt} = \beta_2 \frac{Y_S}{Y_S + \alpha Y_I} X_I. \tag{3}$$

The aphids population grows logistically with the intrinsic birth rate r_1 ($r_1 \in \mathbb{R}_+$) and carrying capacity $\eta(Y_S + Y_I)$, where η ($\eta \in \mathbb{R}_+$) is the quality of the quantity of food that favors the birth of aphids, as shown in Equation (4). Although cereal aphids migrate between different plant species, we are interested in studying what happens within the crop, so we assume a variable carrying capacity. Therefore in the absence of plants, our dynamic system is meaningless.

$$\frac{dX_S}{dt} = r_1(X_S + X_I)\left(1 - \frac{X_S + X_I}{\eta(Y_S + Y_I)}\right). \tag{4}$$

If we consider the process by which aphids acquire the virus, then Equation (4) becomes Equation (5), keeping in mind that now the density of infected plants is calculated as $\alpha Y_I/(Y_S + \alpha Y_I)$ and transmission occurs at a rate β_1 ($\beta_1 \in \mathbb{R}_+$).

$$\frac{dX_S}{dt} = r_1(X_S + X_I)\left(1 - \frac{X_S + X_I}{\eta(Y_S + Y_I)}\right) - \beta_1 \frac{\alpha Y_I}{Y_S + \alpha Y_I} X_S. \tag{5}$$

The infected aphids compartment X_I has an inflow due to virus transmission and an outflow due to the natural death of infected aphids at a rate μ ($\mu \in \mathbb{R}_+$), as shown in Equation (6).

$$\frac{dX_I}{dt} = \beta_1 \frac{\alpha Y_I}{Y_S + \alpha Y_I} X_S - \mu X_I. \tag{6}$$

The above description is summarized in the dynamic system given by Equation (7).

$$\begin{cases} \dfrac{dX_S}{dt} &= r_1(X_S+X_I)\left(1-\dfrac{X_S+X_I}{\eta(Y_S+Y_I)}\right)-\beta_1\dfrac{\alpha Y_I}{Y_S+\alpha Y_I}X_S, \\ \dfrac{dX_I}{dt} &= \beta_1\dfrac{\alpha Y_I}{Y_S+\alpha Y_I}X_S-\mu X_I, \\ \dfrac{dY_S}{dt} &= r_2(Y_S+Y_I)\left(1-\dfrac{Y_S+Y_I}{K}\right)-\beta_2\dfrac{Y_S}{Y_S+\alpha Y_I}X_I, \\ \dfrac{dY_I}{dt} &= \beta_2\dfrac{Y_S}{Y_S+\alpha Y_I}X_I. \end{cases} \tag{7}$$

3. Simulations

The dynamic system given by Equation (7) can be used to make predictions about what will happen under certain conditions, *i.e.*, we can answer what happens if we vary the initial conditions or parameters to study how sensitive output of interest is to an increase or decrease in the amount of other factors. We start the analysis of our system by varying the initial condition since the value of the variables at some point designated as the initial time, typically at $t = 0$, is determinant in the dynamics since this value or initial condition affects the value of the state variables at any future time. Furthermore, according to [16], the crop yield depends on the percentage of aphids and infected plants at the beginning; the simulation parameter values are described in Table 1.

Figure 1 shows a simulation by way of example; we vary the initial condition of susceptible and infected aphids. We observe that as we increase the initial condition, the infected plants become the total population in a shorter time. In Figure 1(a) susceptible plants are extinguished at time $t = 200$, in Figure 1(b) at time $t = 150$, in Figure 1(c) at approximately $t = 110$ and in Figure 1(d) in less than 100. Therefore, considering that the wheat rotation time is 5 to 6 months, we conclude that the initial number of aphids at the start of planting is critical in determining whether all plants will be infected with BYDV until harvest.

Table 1. Simulation parameters.

Parameter	r_1	r_2	η	K	β_1	β_2	α	μ
Value	0.3	0.1	0.5	200	0.1	0.08	0.05	0.1

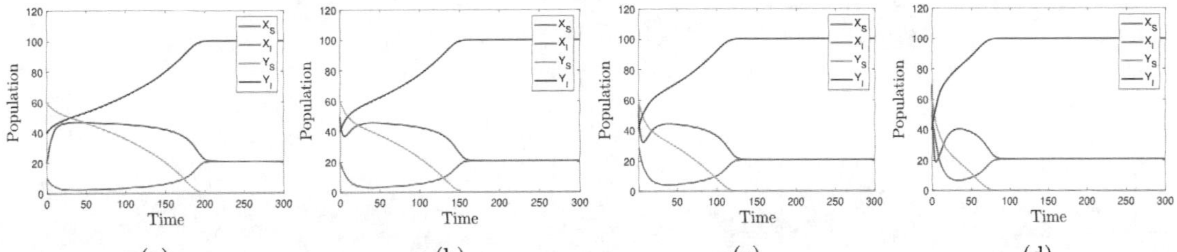

Figure 1. Numerical simulation by varying the initial conditions; (a) $X_S(0) = 20$, $X_I(0) = 10$, $Y_S(0) = 60$ and $Y_I(0) = 40$; (b) $X_S(0) = 50$, $X_I(0) = 20$, $Y_S(0) = 60$ and $Y_I(0) = 40$; (c) $X_S(0) = 70$, $X_I(0) = 30$, $Y_S(0) = 60$ and $Y_I(0) = 40$; (d) $X_S(0) = 70$, $X_I(0) = 50$, $Y_S(0) = 60$ and $Y_I(0) = 40$. The parameter values are described in Table 1.

Figure 2 shows the simulation of the effect of varying the value of eta, which is the quality of the quantity of feed that favors the birth of aphids; we observe that the higher the quality of the food, the more infected aphids reach equilibrium in a shorter time. For example,

Journal of Physics: Conference Series **2515** (2023) 012005 doi:10.1088/1742-6596/2515/1/012005

in Figure 2(a), equilibrium is caught in approximately a time $t = 240$; while in Figure 2(d), approximately equilibrium is reached in time $t = 150$. So modifying food quality for aphids is critical if we want not all plants to be infected until harvest. Therefore, one way to keep the BYDV a threshold is to lower the quality of the aphid food with some chemical application that does not damage the plant and is safe for humans. In fact, in the paper [17], they provide that it is more cost-effective to apply less toxic insecticides for aphids that cause migration to non-crop plantings rather than using products that also kill the natural enemies of aphids.

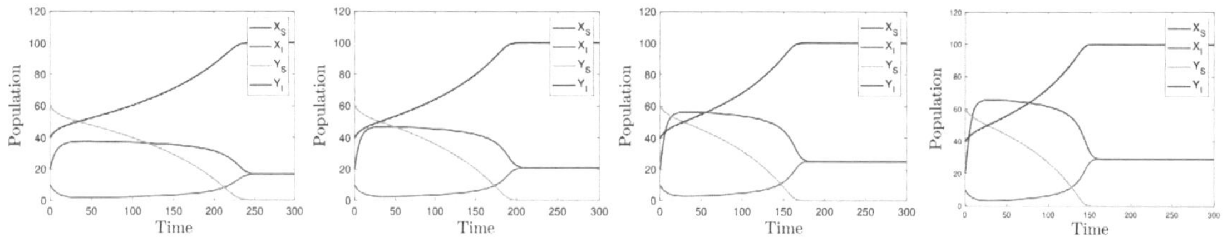

(a) (b) (c) (d)

Figure 2. Numerical simulation varying the quality of the quantity of food that favors the birth of aphids η, and initial condition $X_S(0) = 20$, $X_I(0) = 10$, $Y_S(0) = 60$ and $Y_I(0) = 40$; (a) $\eta = 0.4$, (b) $\eta = 0.5$, (c) $\eta = 0.6$, (d) $\eta = 0.7$. The other parameter values are fixed and are defined in Table 1.

Figure 3 shows the effect of the relative preference of aphids to feed on infected plants; we observe that the lower the preference for infected plants, the lower the number of plants with the virus. Concerning this preference, in the paper [18], they state that pathogens, including the virus, can induce phenotypic changes within the host plant, which leads to the fact that aphids do not act as vectors of random transmission, but feed preferentially on one kind or another depending on the epidemiological state. If we take into account that the rotation time is six months, we observe that in Figure 3(a), we have more plants without viruses at the end of the harvest, while in Figure 3(b), Figure 3(c), and Figure 3(d) all plants will already have virus at harvest. So controlling this preference by some chemical signal would be very helpful to improve crop yield. For example, in [19], they take advantage of knowledge about elicitors that prime the plant immune system against attack to test the effect of a commercial elicitor on infection rates, finding that artificially boosting immunity delayed disease and decreased symptoms.

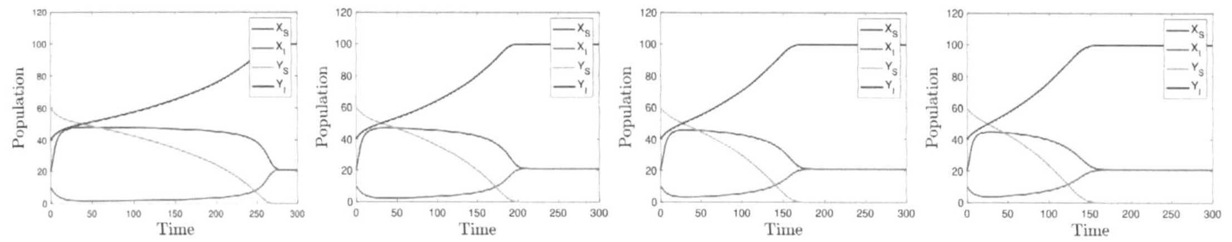

(a) (b) (c) (d)

Figure 3. Numerical simulation by varying the relative preference on aphids to feed of infected plant α; (a) $\alpha = 0.03$, (b) $\alpha = 0.05$, (c) $\alpha = 0.07$, (d) $\alpha = 0.09$. The other parameter values are fixed and are defined in Table 1.

4. Conclusions

Mathematical modeling was used to study the dynamics of aphid-plant interaction and to address the main problem associated with the aphids' preference for colonizing plants with the virus, which leads to the rapid growth of infected plants and a substantial decrease in crop productivity. Numerical simulations allowed us to identify that parameters such as the quality of the food consumed by aphids can directly influence the spread of the virus. To minimize this problem, a control method is proposed that acts in conjunction with rotation times to limit aphid growth until the harvest period. The dynamic system itself has interesting qualitative properties to be further investigated. One such property is that we proposed a variable carrying capacity for aphid growth, which is the maximum growth of any population. In this particular case, we were interested in aphid population growth within the crop. Therefore a plant density-dependent carrying capacity is suitable and provides a more realistic understanding of the dynamics. Considering that most of the research in the literature generally assigned the environmental carrying capacity of a population model as a constant, our model has the potential to generate new qualitative results that can be used to study different phenomena.

Another specific property in our model is the virus transmission function; generally, the law of mass action is used to model transmission. In this aspect, the authors used this function but with aimed to examine the mechanisms regulating pathogen diversity and considered state variables for three viruses, two aphid vectors, and a perennial grass host; in contrast, we incorporated this function into a simple host-vector model that allowed us to accurately examine the effect of aphid preference for feeding on infected plants. It is also noteworthy that the formulation of the dynamic system presented in this paper is a clear example that the future of science is interdisciplinarity; we need knowledge of variations in mathematics and concepts associated with agroforestry systems, ecology, and epidemiology.

Acknowledgments

This work was funded by the Agencia Nacional de Investigación y Desarrollo (ANID)/ scholarship program / 2021-21211843, 2022-21220765. L Cuesta-Herrera, A Altamirano-Fernández and A Vásquez-Coronel thanks to the "doctoral scholarship", awarded by the Vicerrectoría de Investigación y Postgrado at Universidad Católica del Maule, Talca, Chile.

References

[1] Shewry P R and Hey S J 2015 The contribution of wheat to human diet and health *Food and Energy Security* **4(3)** 178–202
[2] Bouis H E 2003 Micronutrient fortification of plants through plant breeding: can it improve nutrition in man at low cost? *Proceedings of the Nutrition Society* **62(2)** 403–411
[3] Aradottir G I and Crespo-Herrera L 2021 Host plant resistance in wheat to barley yellow dwarf viruses and their aphid vectors: a review *Current Opinion in Insect Science* **45(1)** 59–68
[4] Walls J, Rajotte E and Rosa C 2019 The past, present, and future of barley yellow dwarf management *Agriculture* **9(1)** 1–16
[5] Van den Eynde R, Van Leeuwen T and Haesaert G 2020 Identifying drivers of spatio–temporal dynamics in barley yellow dwarf virus epidemiology as a critical factor in disease control *Pest Management Science* **76(8)** 2548–2556
[6] Jarošová J, Żelazny W R and Kundu J K 2019 Patterns and predictions of barley yellow dwarf virus vector migrations in central europe *Plant Disease* **103(8)** 2057–2064
[7] Fitri I, Bakhtiar T, Hanum F and Kusnanto A 2021 Optimal strategy in controlling non-vector pest insect using green insecticide and mating disruption with cost-effectiveness analysis *Journal of Physics: Conference Series* **1796(1)** 012037
[8] McElhany P, Real L A and Power A G 1995 Vector preference and disease dynamics: a study of barley yellow dwarf virus *Ecology* **76(2)** 444–457
[9] Hu Z, Chai R, Liu X, Dong Y, Su D, Desneux N, Tan X and Luo C 2021 Barley yellow dwarf virus–infected wheat plant modulated selection behavior of vector aphids *Journal of Pest Science* **95(3)** 1273–1285

[10] Quillec L L, Plantegenest M, Riault G and Dedryver C 2000 Analyzing and modeling temporal disease progress of barley yellow dwarf virus serotypes in barley fields *Phytopathology* **90(8)** 860–866
[11] Zhang X S and Holt J 2001 Mathematical models of cross protection in the epidemiology of plant–virus diseases *Phytopathology* **91(10)** 924–934
[12] Amaku M, Burattini M N, Coutinho Bezerra F A and Massad E 2010 Modeling the competition between viruses in a complex plant–pathogen system *Phytopathology* **100(10)** 1042–1047
[13] Jeger M, Chen Z, Powell G, Hodge S and Van den Bosch F 2011 Interactions in a host plant–virus–vector–parasitoid system: Modelling the consequences for virus transmission and disease dynamics *Virus Research* **159(2)** 183–193
[14] Amelia R, Anggriani N, Istifadah N and Supriatna A 2021 Stability analysis for yellow virus disease mathematical model of red chili plants *Journal of Physics: Conference Series* **1722(1)** 012043
[15] Mordecai E A, Gross K and Mitchell C E 2016 Within–host niche differences and fitness trade-offs promote coexistence of plant viruses *The American Naturalist* **187(1)** 13–26
[16] Fabre F, Dedryver C A, Leterrier J and Plantegenest M 2003 Aphid abundance on cereals in autumn predicts yield losses caused by barley yellow dwarf virus *Phytopathology* **93(10)** 1217–1222
[17] Jansen J P 2000 A three–year field study on the short–term effects of insecticides used to control cereal aphids on plant–dwelling aphid predators in winter wheat *Pest Management Science: Formerly Pesticide Science* **56(6)** 533–539
[18] Rajabaskar D, Bosque-Pérez N A and Eigenbrode S D 2014 Preference by a virus vector for infected plants is reversed after virus acquisition *Virus Research* **186(1)** 32–37
[19] Kenney J R, Grandmont M E and Mauck K E 2020 Priming melon defenses with acibenzolar-s-methyl attenuates infections by phylogenetically distinct viruses and diminishes vector preferences for infected hosts *Viruses* **12(3)** 1–18

Study of the stiffness and lateral displacements of a confined masonry wall using linear static analysis

L G Echavarría[1], J A Díaz[1], and P A Ospina Henao[1]
[1] Universidad Santo Tomas, Bucaramanga, Colombia

E-mail: paolo.ospina@ustabuca.edu.co

Abstract. Through this investigative study, a computational model will be carried out based on the physical and mechanical properties of a confined masonry wall through specialized software in structures, the above will be done in conjunction with the theoretical study of the finite element method, analysis linear static and its implementation in the model, this will allow a detailed analysis of the stiffness and average lateral displacement of the structural element with areas without discretization, and different sizes of discretization when subjected to a lateral load distributed in the upper part of the structure a slab. This research project will allow the development of a broader perspective on this type of structures and their mechanical behavior when subjected to external disturbances; In addition to providing the possibility of managing critical thinking about decisions to consider in future research on the type of modeling that best behaves in this type of structural elements. The main focus of physics in the article focuses on the need to improve the safety of houses built with adobe against possible seismic phenomena that can cause failures and weaknesses in their structural part, that is, to lay the foundations for the study of the reinforcement of these structures with the contribution of mechanical properties, and it is expected that the mechanical behavior before gravitational loads and accelerations produced by seismic loads in the masonry wall will improve significantly. By means of mathematical models and explained by means of physics, it is expected to estimate the efforts that a wall without reinforcement would present before the loads and compare them with the results obtained in laboratory tests.

1. Introduction

It has been shown throughout history that architecture based on construction processes related to masonry plays an important role in society; In addition, it can be observed that in the municipalities, masonry-based domestic architecture lacks strict compliance with current regulations, and in turn, the use of empirical construction techniques with masonry material prevails [1].

Therefore, the need has arisen to carry out investigative work that allows for a better technical perception of the mechanical and physical properties of this type of element, both when used individually and when used in conjunction with other types of elements and materials [2]. For this reason, the use of computational and mathematical tools is given that allows modeling this type of structural elements in conditions that are as close as possible to those that they would experience, and thus be able to obtain results faithful to the mechanical behavior, which will allow in the future predict failure situations that may affect the integrity and safety of users [3-5].

By means of the theory based on linear static analysis, a computational model is made in SAP 2000 [6] in which the characteristics of the materials, wall sizing and other parameters are described, this to analyze the behavior it has when subjected to a design load previously analyzed in an experimental study; all the above is represented by the average side shift values and structure stiffness. This research article is divided into 4 parts: in the methodology and materials section, all the numerical, computational and parameterization analysis required to carry out the analysis is presented. Section 3 presents the results and provides a brief discussion of each section. And finally in section 5 the conclusions are presented.

2. Methodology and materials

Finite element method is based on discretizing a domain into a small continuum, finding the solution for each one, and then assembling the results. However, this method can be very complex from a mathematical point of view. That is why several requirements were considered for its correct operation, such as that a continuous function under a global domain can be approximated by a series of functions operating under a finite number of smaller subdomains called elements. The points that define the joints between elements are called nodes. These nodal variables refer to degrees of freedom (DOF) and represent geometric domains [7]. In our case, the bridge is a structural problem, and the domain-wide solution sought includes displacement, stress, and strain fields.

The principle of virtual work is the equivalent of stating the equilibrium equations of our system. We consider that each piece of the wall is a solid body that occupies a volume V at an instant of time t and is subjected to forces is $b\,(x,t)$, which act on the contour Γ. These forces produce a field of displacements $u\,(x,t)$, deformations $\epsilon\,(x,t)$ and stresses $\sigma\,(x,t)$, which is described in Equation (1) as evidenced in [8].

$$\int_V \delta\epsilon^T \sigma dV = \int_V \delta u^T bdV + \int_\Gamma \delta u^T t\, Dv\gamma, \tag{1}$$

where the virtual strain field, in Voigt notation, is $\delta\epsilon = \left[\delta\epsilon_x\ \delta\epsilon_y\ \delta\epsilon_z\ \delta\epsilon_{xy}\ \delta\epsilon_{xz}\ \delta\epsilon_{yz}\right]^T$, here $\epsilon_x, \epsilon_y, \epsilon_z$ are the main diagonal elements of the strain tensor and are called normal strains, while $\gamma_{xy}, \gamma_{xz}, \gamma_{yz}$ are the outside the main diagonal components and are called shear stains. In addition, the forces of the bridge are presented in the expression $b = \left[b_x\ b_y\ b_z\right]$. In addition, the surface forces acting on the contour of the volume of each piece of the structure are considered. They are produced by the contact of the parts with their exterior where they are located.

The traction vector with three components in the three orthogonal directions is presented as $t = [t_x\ t_y\ t_z]$. Finally, in Equation (1) the stress term is described, and we consider that the mass and surface forces act on components. We can present the Cauchy stress tensor for a differential volume. Assuming a linear elastic constitutive law, stress and strains are calculated with the Equation (2) as evidenced in [2].

$$\sigma_{ij} = \frac{E}{(1+v)}\left[\epsilon_{ij} + \frac{v}{(1-2v)}\delta_{ij}\,\epsilon_{kk}\right]. \tag{2}$$

In Equation (2), E is elastic modulus, and υ is Poisson's modulus. The terms $\sigma_x, \sigma_y, \sigma_z$, are the main diagonal elements of the stress tensor and are called normal stresses, $\tau_{xy}, \tau_{xz}, \tau_{yz}$, are the components outside the main diagonal and are shear stress which is composed of in-plane and out-of-plane stress. Moreover, an equivalent acting stress needs to be computed when there is more than one stress component. For that, the Von Mises criterion is described by Equation (3) as evidenced in [2]; σ_i is the normal stress parallel to each of the three principal axes, N is the factor of safety, and σ_f is the elastic limit.

$$\frac{\sigma_f}{N} = \sqrt{\frac{(\sigma_1-\sigma_2)^2+(\sigma_2-\sigma_3)^2+(\sigma_3-\sigma_1)^2}{2}}. \tag{3}$$

On the other hand, if a component is subjected to fluctuating stresses, to predict life, the stress-life method, σN_f , can be used, where σ_f is stress and N_f number of cycles to failure [9]. Their relation is described by the Basquin rule, which is shown in Equation (4) as evidenced in [9].

$$N_f = A\sigma_f^m, \tag{4}$$

where A and m are material constants. Furthermore, the Miner-Palmgreen damage rule is the simplest but most widely used accumulated damage model [4]. It says that if there are different events c_k at a stress σ_k and the average number of cycles N_i to produce failure at a stress σ_i, then the damage fraction C given by an event k is given by the Equation (5) as evidenced in [2], which represents the sum of the individual contribution of the product stress and cycles in each event k. Then, C is the fraction of life consumed by exposure to k cycles at different stress levels to which the bridge is subjected.

$$C = \sum_{k=1}^{i} \frac{\sigma_k \eta_k}{\sigma_i N_i} \leq 1. \tag{5}$$

To carry out the calibration of the model, some initial data are needed and then some results from which the comparison is made [10]. In this case, as initial data, the physical and mechanical properties of the masonry were required (which will be modeled as an isotropic material and together with the glue mortar to simplify the modeling and be consistent with the undergraduate knowledge of the authors), and the concrete with which the confinement elements used to improve ductility are represented, as shown in Table 1 [11].

For the results that are required for the subsequent comparison with those obtained in the numerical modeling, the values in Table 2 are available; all these data were obtained from the experimental results published by López [12]. The results from which the comparisons with the experimental test are made are those of stiffness (t/mm) Equation (6) from a lateral load of 5.2 tons; this expression relates the experimental lateral load to the average displacement obtained in the deformed model.

$$K_{exp} = \frac{V}{\delta}. \tag{6}$$

Table 1. Mechanical and physical properties of materials [12].

Properties	Masonry	Concrete
Compressive strength (kg/cm^2)	32.60	154.70
Elasticity modulus (kg/cm^2)	19579.50	99502.30
Shear modulus (kg/cm^2)	3915.90	41459.30
Poisson's coefficient	0.25	0.20
Volumetric weight (kg/cm^3)	1300	2400

Table 2. Stiffness and average lateral displacement of the slab for a lateral load of 5.2 tons experimental wall [12].

-	Average lateral displacement (mm)	Stiffness (t/mm)
Experimental wall	0.84	≈6.19

On the other hand, it should be clarified that the modeling was carried out by the linear static analysis method to simplify the process and to be able in future research to compare the results with a more complex non-linear method and thus calculate the percentage of error that was found. obtained by doing it by both processes (see Figure 1) [13].

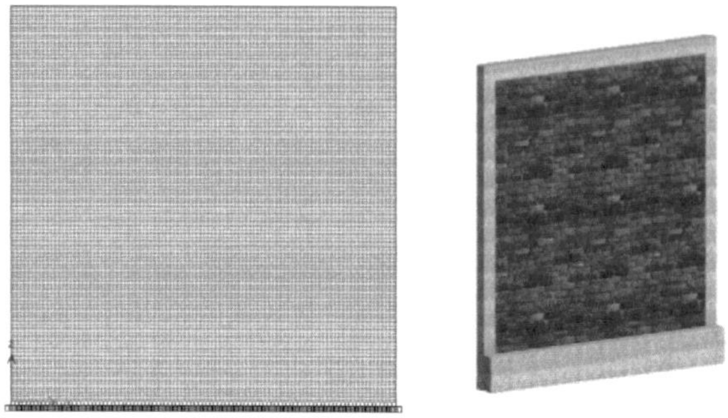

(a) (b)

Figure 1. Masonry wall: (a) Numerical model in SAP 2000 [6], and (b) Rendered model in Revit [14].

3. Results and discussions

This section presents the analysis of the results obtained from the modeling in SAP 2000 [6] of confined masonry walls in comparison with those obtained from tests carried out in the laboratory with the same materials.

3.1. Slab displacements and stiffness from applied load in a non-discretized masonry wall

Next, in Figure 2 we can observe the deformed model of the confined masonry wall without discretizing when applying a lateral load of approximately 5.20 tons, in addition, it can be evidenced in Table 3 the stiffness and the average lateral displacement of the slab that is found. located at the top of the element; said displacement was obtained by averaging the value of three different places, in the two corners and in a central point [4,15,16].

Figure 2. Deformed model of a masonry wall without discretization.

Table 3. Stiffness values and average lateral displacement of the slab for a lateral load of 5.2 tons.

-	Average lateral displacement (mm)	Stiffness (t/mm)
Non-discretized wall	1.352	≈ 3.840

3.2. Slab displacements from the applied load in a discretized masonry wall in 10×10 areas

On the other hand, in Figure 3 we can observe the deformed model of the discretized confined masonry wall in 10×10 areas when applying a lateral load of approximately 5.20 tons, in addition, it can be seen in Table 4 the average lateral displacement of the slab that is located at the top of the element; said displacement was obtained by averaging the value of three different places, in the two corners and in a central point [4,15,16].

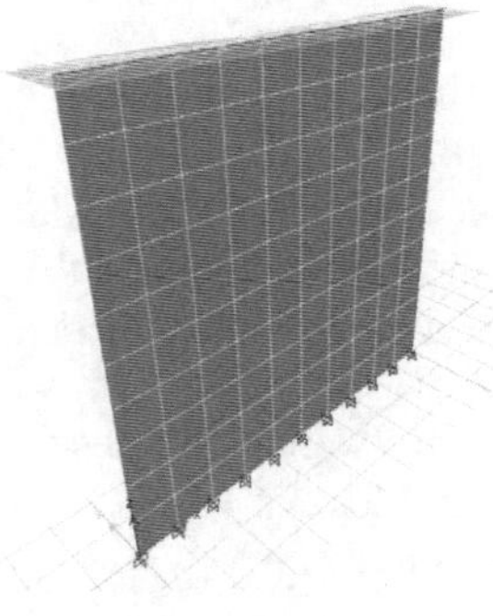

Figure 3. Deformed model of a discretized masonry wall.

Table 4. Stiffness values and average lateral displacement of the slab for a lateral load of 5.2 tons.

-	Average lateral displacement (mm)	Stiffness (t/mm)
Discretized wall 10×10	1.571	≈3.300

3.3. Displacements of the slab from the load applied on a discretized masonry wall 100×100

Now, in Figure 4 we can see the deformed model of the confined masonry wall discretized in 100×100 areas when applying a lateral load of 5.20 tons approximately, Table 5 also shows the average lateral displacement of the slab located at the top of the element; this displacement was obtained by averaging the value of three different places, in the two corners and at a central point [4,15,16].

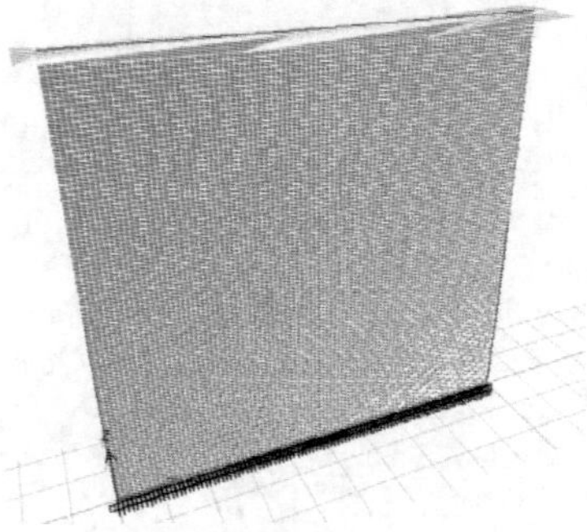

Figure 4. Deformed model of a discretized masonry wall in 100×100 areas.

Table 5. Stiffness values and average lateral displacement of the slab for a lateral load of 5.2 tons.

-	Average lateral displacement (mm)	Stiffness (t/mm)
Discretized wall 100×100	1.458	≈ 3.560

3.4. Results comparison

In Table 6, we can see the results obtained from the computational modeling in SAP 2000 [6] of confined masonry walls by the linear static analysis method in comparison with those obtained from the same element tested through an experimental method.

Table 6. Results obtained from the computational modeling and experimental test of the confined masonry wall [12].

-	Average lateral displacement (mm)	Stiffness (t/mm)
Non-discretized wall	1.352	≈ 3.84
Discretized wall 10×10	1.458	≈ 3.56
Discretized wall 100×100	1.571	≈ 3.30
Experimental test wall	0.840	≈ 6.19

4. Conclusions

Firstly, by means of the results obtained from the modeling of the structural element in the SAP 2000 software, we can conclude that the linear static analysis method generates data that is far from that obtained in the experimental test, even handling the same properties of the materials, this may be due to the fact that they only allow one data to be obtained for the stiffness and displacements in (which differs from the experimental behavior in which several displacement data are presented, and therefore of stiffness). On the other hand, we can observe that the average displacement values in the experimental test are higher, this may be due to the way in which the software estimates the stiffness (lower values due to material degradation) compared to the experimental test. in which the stiffness is much higher, which may be due to the properties of the material in its initial state.

Acknowledgments

We are grateful for the permanent support of the Vicerrectoría Académica of Universidad Santo Tomas.

References

[1] Sánchez E 2007 La arquitectura de tierra en Colombia, procesos y culturas constructivas *Apuntes* **20(2)** 242
[2] Juárez G, Ortiz P 2021 Importancia del mallado de elementos finitos de muros de mampostería confinada en el análisis sísmico de edificios *Revista Internacional De Ingeniería De Estructuras* **26(4)** 712
[3] Sánchez S, Mebarki A 2009 Método semi-empírico para estimar la resistencia lateral en muros de mampostería confinada *Ingeniería Sísmica* **80** 113
[4] Cornejo T, Hernández, H 2013 Resistencia lateral de estructuras de mampostería confinada *Ingeniería* **17(3)** 179
[5] Zúñiga O, Terán A 2008 Evaluación basada en desplazamiento de edificaciones de mampostería confinada *Ingeniería Sísmica* **79** 25
[6] CSI Spain 2023 *CSI Bridge* (Madrid: CSI Spain)
[7] Barbero E 2013 *Finite Element Analysis of Composite Materials Using ANSYS* (Boca Raton: CRC Press)
[8] Kurowski P 2018 *Engineering Analysis with Solidworks Simulation* 2018 (Mission: SDC Publications)
[9] Ashby M 2011 *Materials Selection in Mechanical Design* (Oxford: Elsevier)
[10] López C 2017 *Modelado no lineal de Muros de Mampostería Combinada y Confinada* (Ciudad de México: Universidad Autónoma Metropolitana Unidad Azcapotzalco)
[11] Ochoa M, Quiñones K 2020 *Elaborar una Guía Práctica para el Cálculo, Diseño y Control de Muros de Contención en Mampostería Confinada y Concreto Reforzado en Voladizo, Calibrando Resultados con el Software SAP 2000* (Ocaña: Universidad Francisco de Paula Santander)
[12] López X 2021 *Modelado no Lineal de un Edificio de Tres Niveles a Base de Muros de Mampostería* (Ciudad de México: Universidad Nacional Autónoma de México)
[13] Orduña A, Ayala G 2001 Análisis no lineal de muros de mampostería confinada sujetos a cargas laterales *Revista Engenharia Civil* **11** 21
[14] Autodesk, Inc. 2022 *Autodesk Revit* (United States of America: Autodesk, Inc.)
[15] López J, Oller S, Oñate E 1998 *Cálculo del Comportamiento de la Mampostería Mediante Elementos Finitos* (Barcelona: Centro Internacional de Métodos Numéricos en Ingeniería)
[16] Bonilla D, Merino G, Sosa D 2017 Study of bamboo physical properties and its application as reinforcement in adobe structures *International Review of Civil Engineering* **8(4)** 160

Analysis of the stresses for a bridge by the finite element method under the action of viscous damping

R D Oñate Ballesteros[1], L A Balcázar[1], A Navarro[1], and P A Ospina Henao[1]
[1] Universidad Santo Tomás, Bucaramanga, Colombia

E-mail: paolo.ospina@ustabuca.edu.co

Abstract. Specifically speaking of the treatment of constructions (such as bridges), the importance of considering dynamic phenomena from the design phase lies in the fact that the protection of the construction against unwanted external disturbances, such as those produced by earthquakes and vehicle traffic, is guaranteed. automotive, among other causes, that may cause partial or total damage to the structure. As a tool, we have computer-aided design and engineering for the structuring of mechanical designs, which adds efficiency to detect design errors in early phases. The relevance of this project lies in the application of knowledge to the areas of computer-aided design in civil structures such as bridges, through computerized simulation in CSI Bridge and mathematical analysis to determine the efforts and graphically observe the behavior of the bridge when subjected to viscous damping. The physical relevance of this project lies in the application of knowledge to the design areas of civil structures such as bridges, through computerized simulation and mathematical and physical analysis to determine the efforts and graphically observe the behavior of the bridge when it is subjected to external disturbances that act as viscous damping. Physically, we model this damping as a friction force in general to solve the equation using numerical methods by changing different boundary conditions in the damping factor, which adds relevance to the computational modeling of the bridge and the study of its physical and mechanical properties.

1. Introduction

At present it is impossible to imagine a good quality of life for the inhabitants of different societies without thinking about the use of different structures; Being these parts of the bases of our civilization, through the years, different studies and investigations have been carried out, in addition to improving and implementing different techniques that guarantee the security and integrity of the users who use it.

This is the case of bridges, which allow us to move more easily through areas where it would be difficult and even impossible to travel without them; and in order to guarantee that their functionality lasts over time, regulations have been created throughout the world, in the case of Colombia, the Ministerio de Transporte and the Instituto Nacional de Vías (INVIAS) in 2013 in agreement with the Asociación Colombiana de Ingeniería Sísmica (AIS), made the Norma Colombiana de Diseño de Puentes CCP-2014 [1], which through probabilistic techniques and other tools allow a safer and more effective design.

On the other hand, a fundamental theory in structural design philosophy has been developed; reliability, which allows creating, evaluating, and designing computational and mathematical models of these structures considering factors such as dynamic loads, in this case the load represented by vehicles that constantly circulate and in which their mass varies [1].

Taking into account the concepts of structural engineering that apply to the case, a model is proposed in CSI Bridge [2] in order to describe the behavior and stresses presented by the bridge designed before the loads of the pavement to which it is permanently subjected, the above is represented by means of the stress diagrams in X direction and Y direction of the slab; In addition, a section is made with a numerical analysis carried out with a mathematical method that allows to observe by means of diagrams the behavior of the structure when it presents an external damping, in this case by means of a viscous damper. The following writing is distributed in the following sections: Section 2 presents all the methodological part, equations and properties of the materials used in modeling at CSI Bridge [2]. The following section presents the results of modelling and numerical analysis of the structure subjected to viscous damping. Finally, the conclusions are presented in section 5.

2. Methodology and materials

For the analysis of structural dynamics and viscous damping in a complex structure such as a bridge, it is important to take into account all the variables and technical factors that may affect the quality of the results obtained; computational and mathematical analysis are the bases for this type of research, our case in particular, required the handling of both experimental data through laboratory tests, as well as the computational analysis of the equations that allowed comparing the results. In the part of computational modeling, the correct handling of materials and their physical and mechanical properties is fundamental, to obtain a model with a behavior that is closest to reality, and therefore also its results. The foundations that allowed this project to be carried out are presented below.

2.1. Bridge subjected to viscous damping

Due to its analytical simplifications, the most used type of damping in structural dynamics is viscous damping [3]. This does not imply that the real damping found in structural systems is always of this type, but rather it is based on the idealization of being able to satisfactorily represent the vibrational behaviour of real structures with minimal mathematical implications [4]. In our case, the study of a one degree of freedom structure with a damping force can be idealized as Figure 1 (a) and modeled in CSI Bridge [2] as seen in Figure 1 (b).

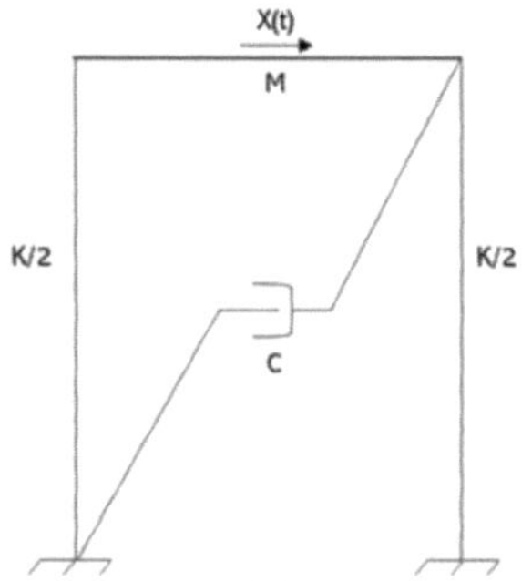

(a)

(b)

Figure 1. One degree of freedom structure with a damping force; (a) ideal single degree of freedom system with viscous damping; and (b) bridge designed in CSI Bridge [2].

In realistic situations, the structure is inevitably losing energy due to frictional forces and consequently its amplitude decreases with time [5]. To explain the gradual loss of energy, different types of dissipative forces can be incorporated into the system. However, we will assume a nonconservative force proportional to a power of the velocity, this generalized friction force can be expressed as Equation (1) [6].

$$F_{\text{frict}} = \begin{cases} -\gamma\,\dot{x}^{n}, \text{for odd n} \\ -\gamma\,\text{sgn}\,(\dot{x})\dot{x}^{n}, \text{for even n} \end{cases}, \tag{1}$$

where n is assumed to be a positive integer exponent and sgn is the sign or signum function that takes the values 1, 0, −1 for $\dot{x} > 0$, $\dot{x} < 0$ and $\dot{x} = 0$, respectively. Since the frictional force is always directed against the direction of the velocity, the use of the sgn function is very important to obtain an adequate model of this force. In this way, the position, and the velocity change correctly from positive to negative (and vice versa) when the bridge passes through the point where it stabilizes. Now, both cases can be summarized in one equation by introducing a new parameter ε_n defined by Equation (2) [6].

$$\varepsilon_n = \begin{cases} 1, \text{for odd n} \\ \text{sgn}\,(\dot{x}), \text{for even n} \end{cases}. \tag{2}$$

If substituting Equation (2) in Equation (1) which are written as [6], we obtain a more general expression for the friction force, that is $F_{\text{frict}} = -\gamma\varepsilon_n\dot{x}^n = -\dot{\gamma}\dot{x}|\dot{x}|^{n-1}$, which works in the case of integer exponents, however, n can be any arbitrary index, that is, it can be assumed to be a non-integer power, but such situations are not experimentally motivated and will not be analyzed in this article. Furthermore, it is numerically cheaper to use the function ε_n than the product $\dot{x}|\dot{x}|^{n-1}$ since the factor ε_n can be entered as a helper int function that returns a constant value in C++ [7] or Matlab [8]. So, using classical mechanics, the equation of motion has the form of Equation (3).

$$\ddot{x} + \gamma\varepsilon_n\dot{x}^n + \omega^2 = 0, \tag{3}$$

where, $\gamma = C/M$ represents the damping constant of the system due to the friction force and $\omega^2 = K/M$, represents the natural angular frequency of oscillation. Both physical properties depend on the mass of the bridge, the self-damping, and the stiffness of the structure [9].

2.2. *Initial bridge design, materials, and loads*

The initial design of the bridge and its basic elements considered the pre-dimensioning of the structural elements along the element, the general geometry in terms of plan and longitudinal profile, geometry of details such as traffic routes and protective elements as seen in Figure 2(a), Figure 2(b), and other important characteristics according to the parameters stipulated in the specifications of the codes of design and construction of bridges of the Colombian regulation of bridge design[1,10].

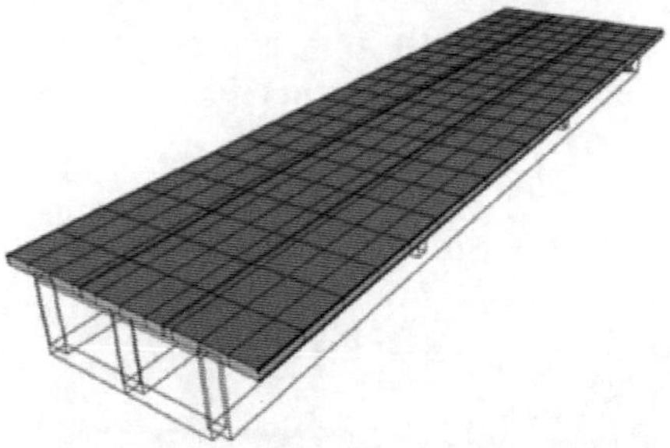

(a)

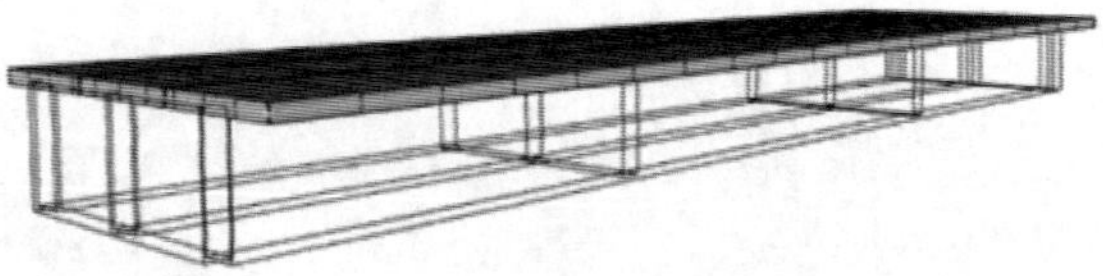

(b)

Figure 2. Three-dimensional view of the bridge modeled in CSI Bridge [2]; (a) bridge slab, and (b) beam elements.

An important aspect to obtain a modeling of structures that resemble their properties, characteristics and behavior that would occur in a real event, are the materials and loads that compose them; in this case the physical and mechanical properties of materials such as 28 MPa concrete and 420 MPa steel were considered. As for the loads was performed the analysis with 2 types; on the one hand, we have the weight of the pavement (DW), which corresponds to the weight of this element that composes the bridge and is distributed along the entire slab, and on the other hand we have the vehicular live load (LL) that corresponds to a dynamic weight in motion that is stipulated according to the specifications of the codes of design and construction of bridges of the Colombian regulation of design of bridges [1]. In this order of ideas, the Table 1 and Table 2 are presented with the data with the properties of concrete, steel and load values, characteristics that were considered to obtain the results of the efforts presented by the elements for each type of load mentioned above [11].

Table 1. Mechanical and physical properties of concrete.

Properties	Magnitude	Units
Compressive strength	28	MPa
Tensile strength	3.28	MPa
Elasticity modulus	24855.578	MPa
Poisson's coefficient	0.20	-
Specific weight	24	KN/m^3

Table 2. Load values for analysis.

-	KN/m^2	Vehicle typle	KN/m
Load weight of the pavement (DW)	1.5	-	-
Live vehicle cargo (LL)	-	W14	10.3

3. Results and discussions

Next, the data resulting from the modeling of the bridge in the CSI Bridge [2] software are presented, in which the efforts generated by the assignment of the pavement load to the slab can be highlighted, in addition to the results of the mathematical modeling of the structure at be subjected to viscous damping to conclude the research study.

3.1. Numerical analysis for the underdamper case

The numerical study of Equation (3) is presented for the underdamped case, assuming a constant damping coefficient and angular frequency, since these values are fixed and specific to the physical characteristics of the bridge, also considering even and odd powers for n. A 4th order Runge-Kutta (RK4) was implemented to integrate the system [12]. Now, to implement the numerical method, we write Equation (3) as a system of ordinary differential equations introducing the following change of variables, these are $x_0 = x$ and $x_1 = \dot{x}$; alternatively, we introduce a vector field F, and write the system as follows $x = [x_0, x_1]$ and the vector field $F = [\dot{x}_1, -\gamma\varepsilon_n x_1^n - \omega^2 x_0]$. Numerical solutions are found assuming the following initial conditions $x_0(0) = 2$, $x_1(0) = 0$, and parameters $M = 186{,}485.7$ kg, $\gamma = 0.5$. The Figure 3 shows the bridge displacement for different n [9].

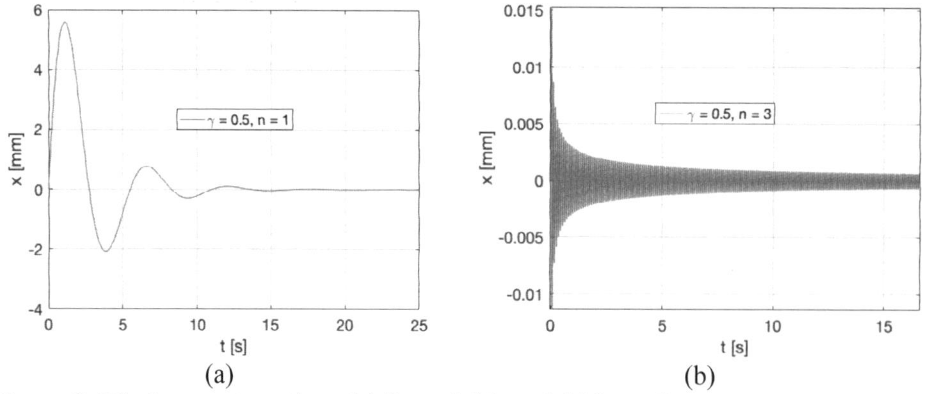

(a) (b)

Figure 3. Displacement vs. time; (a) for n=1 (a), and (b)for n=3.

Oscillations for n = 1 tend to zero very quickly but for n > 1 they fall to zero very slowly as can be seen in Figures 2(a), Figure 2(b). In addition, it is observed that as n takes higher values, the oscillation amplitudes are greater, which are a case study of numerical interest but in reality, it is very difficult to present these phenomena on the structure, for that reason in general the friction forces are taken proportional to the speed, however, you can notice the good performance of the damping factor [13].

3.2. Axial forces in the X direction and Y direction to support the weight of the pavement on the slab
Below, we present in Figure 4(a), Figure 4(b) the stress diagrams for the X and Y directions respectively, obtained by evaluating the weight load of the pavement distributed along the slab; later in Table 3 we can observe the maximum stress values for each direction expressed in (KN/m^2) [14,15].

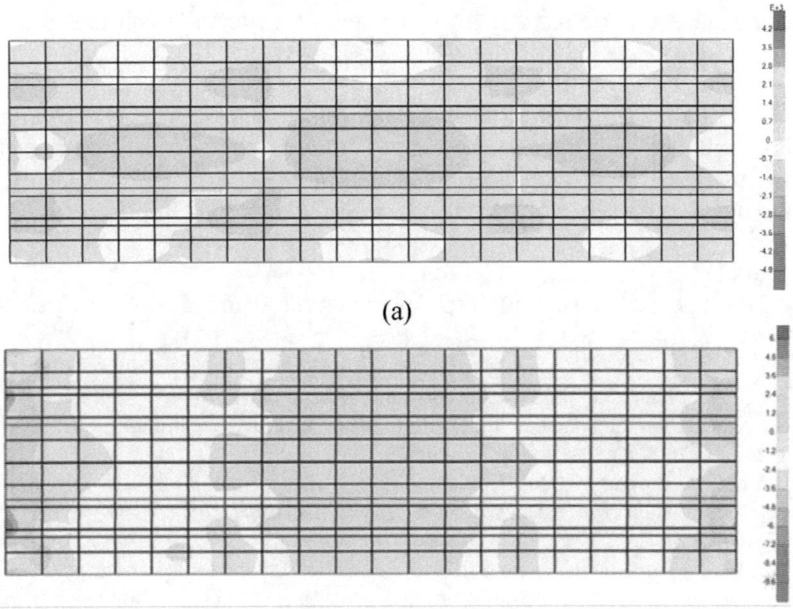

(a)

(b)

Figure 4. Plant view of the forces for the load of the pavement weight on the slab (KN/m^2); (a) X direction, and (b) Y direction.

Table 3. Minimum and maximum efforts before the load of the weight of the pavement on the slab.

-	Minimum (KN/m^2)	Maximum (KN/m^2)
X direction	≈ -5224.03	≈ 2766.79
Y direction	≈ -10478.55	≈ 3688.75

3.3. Service limit state review
This section presents the results of the serviceability limit state analysis calculated based on the ultimate stresses in compression and tension; values which allow defining a percentage of failure in the structure, and depending on it, stipulate whether the operation will be adequate and guarantee the integrity and safety of the users. These values are expressed in KN/m^2 as shown in Table 4 [13-15].

Table 4. Results of the service limit state analysis (KN/m^2).

-	Allowable stresses (KN/m^2)
Compressive strength	18200
Tensile strength	4528

4. Conclusions

A more general damping model was introduced to analyze how different damping powers affect the dynamics of the bridge system. This model considers an integer exponent n in the friction force of the differential equation and, in the case of even powers, includes the signum function to adequately model the average drag. We compute some numerical solutions of the generalized differential equation in the non-harmonic regime assuming odd powers, since even higher-order ones did not yield significant results.

Numerical solutions were found by implementing the traditional Runge-Kutta solver. In addition, the damping coefficient was calculated with the real values of the bridge dimensions, and it is not large enough, the curves show fluctuations with low amplitudes, that is, small oscillations around the equilibrium point, until equilibrium is reached. We conclude that for $n = 1$, that is, the linear power, the decay is faster to the equilibrium position when γ after a time remains constant, the trend is the same for odd powers.

Finally, considering the work stress values obtained due to the presence of permanent loads such as self-weight along the slab, it can be concluded that it meets the allowable stress requirements along the entire span; this according to the dimensions of the structure, capacity and both physical and mechanical properties of the materials used in the design, which provide considerable support so that current regulatory requirements and stress requirements are met in the most appropriate way.

Acknowledgments

We are grateful for the permanent support of the Vicerrectoría Académica of Universidad Santo Tomás.

References

[1] Asociación Colombiana de Ingeniería Sísmica (AIS) 2014 *Norma Colombiana de Diseño de Puentes, CCP 14* (Colombia: Asociación Colombiana de Ingeniería Sísmica)
[2] CSI Spain 2023 *CSI Bridge* (Madrid: CSI Spain)
[3] Stuardi J, Massa J, Brewer A 2005 Eficiencia de sistemas de aislamiento sísmico basados en resortes helicoidales y amortiguadores viscosos *Revista Internacional de Desastres Naturales, Accidentes e Infraestructura Civil* **7(2-3)** 177
[4] Chopra A 2014 *Dinámica de Estructuras* (México: Pearson Educación)
[5] Aguilar R, Gudiño E 2011 dos métodos simplificados para el análisis sísmico de puentes simplemente apoyados *Revista Ciencia* **14(3)** 1
[6] Quispe J 2021 *Diseño Basado en el Movimiento de Amortiguadores Viscosos para Puentes Rectos* (Madrid: Universidad Politécnica de Madrid)
[7] ISO/IEC 2020 *C++20* (United States of America: ISO/IEC)
[8] The MathWorks 2021 *MATLAB, R2021a* (United States of America: The MathWorks, Inc.)
[9] Samartín A 1968 *Una Aplicación de los Métodos Matriciales al Cálculo de Puentes* (Madrid: Alameda)
[10] Brockenbrough R, Merrit F 1999 *Structural Steel Designer's Handbook* (New York: Mc Graw Hill)
[11] Gómez R 2008 *Comparación Entre las Filosofías de Diseño por Esfuerzos Admisibles, Factores de Carga y Factores de Carga y Resistencia en el Diseño de Superestructuras de Puentes Vehiculares Bajo la Acción de Cargas Gravitacionales* (México: Universidad Nacional Autónoma de México)
[12] Butcher J 2008 *Numerical Methods for Ordinary Differential Equations* (Chichester: John Wiley & Sons)
[13] Arroyo J 2004 Metodología para calcular la respuesta de estructuras con amortiguamiento viscosos *Revista Internacional de Desastres Naturales, Accidentes e Infraestructura Civil* **4(2)** 117
[14] Bonilla D, Merino G, Caiza D 2017 Study of bamboo physical properties and its application as reinforcement in adobe structures *International Review of Civil Engineering* **8(4)** 160
[15] Pérez T, Martínez A 2016 *Criterios de Dimensionamiento en Estado Límite de Servicio de Tableros de Puentes con Sistema de Atirantamiento Extradosado* (Madrid: Universidad Politécnica de Madrid)

Analysis indicators of heart rate variability, before and after active standing with fractional Brownian movement and t-test

M Mejía[1], G Morantes[2], F Rojas[3], E Prada[1], and D Prada[1]
[1] Universidad Pontificia Bolivariana, Bucaramanga, Colombia
[2] Universidad Autónoma de Baja California, Mexicali, México
[3] Universidad Autónoma del Estado de Morelos, Cuernavaca, México

E-mail: duwamg.prada@upb.edu.co, manuel.mejiao@upb.edu.co

Abstract. Alzheimer's is a cognitive neurodegenerative disease that mainly affects older adults. Its detection is generally evident when the disease is in a medium-advanced state of development. The studies are carried out with greater emphasis on the part of the brain, however under scientific evidence there is a direct relationship with psychophysiological measures. Identifying a variable that can evidence the presence of this type of dementia is the purpose of this study. Using fractional Brownian movement and the repeated measures t-test applied to electrocardiographic signals, recorded under the standing task, of 8 patients with the presence of the disease, the change in the maximum and minimum heart rate and the average heart rate were evidenced. the same. With the above, there are indicators of the presence of the disease, which contribute significantly to its diagnosis.

1. Introduction

There are several neurological diseases that have been presenting in adults and even more so in older adults. Some of these diseases have no cure, however there are medical interventions that can improve the quality of life of the patient and his family. The detection of these diseases is generally evident at times when it is already in the middle-advanced or advanced stages, which requires faster intervention, but it is difficult to guarantee quality of life due to the deterioration that is observed. Dementia is a syndrome characterized by cognitive impairment that generates a gradual loss that affects the overall functioning of the person and involves behavioral and psychological symptoms [1]. However, one of the most prevalent dementias is Alzheimer's disease (AD), a neurocognitive disorder that causes a progressive and degenerative deterioration that affects between 60% and 70% of cases witha diagnosis of dementia, which is one of the main consequences of disability in older adults worldwide[2].

The psychological approach to dementia due to conditions such as Alzheimer's disease, which often focuses on assessing the cognitive, behavioral, and emotional changes associated with the impairment, is appropriate considering the pervasive impact these changes have on functionality, autonomy and well-being of the person involved, in addition to being able to be identified by methods such as neuropsychological evaluation combined with clinical evaluation [3]. Likewise, with the same psychological approach, little attention is paid to the physical changes associated with these states, and although many of them focus on brain activity (mainly cortical), it is intended to provide an explanation for these perceptual, behavioral, and emotional changes, less importantly to peripheral physical indicators such as cardiac, vascular and electro dermal activity and their relationship with the above changes and even with the progression of dementia [4].

In relation to the diagnosis of Alzheimer's disease, various computational tools have been created with techniques based on mathematical models that allow experts to contribute to the diagnosis decision. Generally, neuropsychological tests are carried out based on endorsed and standardized instruments that allow a good diagnosis, among the most used is the mini-mental, Montreal cognitive assessment (MoCA). In order to address how autonomic regulation is executed and its association with cognitive functioning, it is necessary to point out that processes such as cardiac, vascular, respiratory and electro dermal activation are controlled by the autonomic nervous system, and to discuss the function of these processes, the literature generally shows control exerted through the interaction of our sympathetic andparasympathetic autonomic branches [5].

In this regard, it should be noted that several autonomic nervecenters associated with this control are anatomically located outside the cerebral cortex and even in brain regions such as the forebrain, for example, it is known that the heart has nodules that manage to carry out action potentials in an efficient manner. which spread throughout the myocardium, efferent control of cardiac and vascular activity is linked to brainstem nerve centers such as the vague nucleus, of which about 75% projects parasympathetic or reticular fibers from the medulla and bulge responsible for central control of vascular tone [6].

The importance of these autonomic centers in the hindbrain is based on their functional relationship with structures such as the spinal cord and diencephalon, the hypothalamus, the amygdala, the septum pellucidum, and brain areas such as the orbitofrontal area and the hippocampus, given that their connectivity influences in homeostasis, the autonomic and hormonal response to stress and emotions [6], given this, fluctuations in the function of the autonomic nervous system may therefore underlie levels of homeostasis, mood, stress level and even cognitive well-being [7].

The above described becomes relevant when it shows that in the last decade, there has been an increase in attention and tastetowards the study of neurological changes in the nucleus located in the brainstem derived from AD, associated with abrupt changes in the levels of serotonin and acetylcholine, whichis why behavioral symptoms associated with mood, emotional reactivity, agitation or alterations in the sleep cycle take on more meaning, given this, in [8], the possibility has been proposed that Alzheimer'sdisease begins with changes in the brainstem, manifested from the beginning of the first phase, through non-cognitive symptoms and even autonomic alterations, which have been omitted in clinical practices,due to diagnostic criteria that are have cataloged as the main dementias, which focus on cognitive deficits, which are assumed to be derived from cortical alterations ales.

To contribute to the diagnosis, the electrocardiographic signal of 8 of the patients diagnosed with Alzheimer's disease was analyzed in this work, using the rescaled range methodology to calculate the Hurst coefficient related to fractional Brownian motion (FBM). [9], which in addition to being used for the study of the dynamics of a particle in a fluid, has been useful in the study of polymer dynamics, seismology, pseudorandom sequences, among others [10] and therefore, it is used in observing persistence in the pure signal.

These results are contrasted with the standing methodology that is generally carried out to study this phenomenon. An FBM with parameter of Hurst $H > 1/2$ is called a persistent process, that is, the increments of this process are positively correlated. Instead, increments of an FBM with $H < 1/2$ they constitute what is called an anti-persistent process, the increases being negatively correlated; for $H = 1/2$, an FBM corresponds to a Brownian motion that has independent increments [11]. From the above, it is intended to analyze a possible relationship between AD through the analysis of the H index and the analysis of heart rate (HR).

2. Mathematical method

Initially, this section describes the materials used and the protocol established for recording the data regarding the standing test. The mathematical method used for the analysis of the collected data is shown below.

2.1. Electrocardiographic signal recording

To carry out the analysis of the psychophysiological records (electrocardiogram), the equipment PowerLab PL 3508 – 8/35 (ADInstruments®). Cardiac activity recording was performed by arranging disposable surface electrodes using the second lead of the electrocardiogram (ECG), these electrodes were made up of positive (+), negative (-) and ground, the configuration had a bandwidth of 0.5 Hz to 150 Hz. HR variability indicators were obtained from this record. Additionally, a mark button was used that was manipulated by the evaluator, to generate precise marks in the physiological records associated with the presentation of the stimuli and/or specific moments. The sampling frequencyfor the recording of all the signals corresponded to 200 Hz.

For data collection, the participant was seated in a completely comfortable, calm and silent position, in the first instance, viewing a completely black screen (initial baseline), this in order to identify their baseline of cardiac contractions over time, then a series of visual and auditory stimuli were presented that would generate significant changes in the participant's HR, once the presentation of the stimuliwas completed, the participant entered a period called final basal, in which is intended to return to its baseline again visualizing the totally black stimulus; after an approximate time of three minutes, the participant was asked to get up from the chair as quickly as possible and remain standing for two minutes, an action that represented an orthostatic tolerance test, known as active standing [12].

2.2. H-index

This section describes the mathematical method used for the analysis of the electrocardiographic signal. Obtaining the data to be processed in this work is derived from one minute prior to the order to get up (pre) and one minute after getting up (post), to identify possible changes in the HR of each of the participants. With these pre and post signals, the Hurst (H) coefficient calculation procedure is performed on each one to observe the state before the test and after the test. To determine the value of H specifically, a partition of the time series is performed in subgroups, with each of these subgroups the standard deviation and the rescaled range is calculated, which is the result of the difference between themaximum value and the minimum value of cumulative deviations.

Next, a linear regression is performedbetween the values of the natural logarithm (Ln) of the data number by subgroup (N_i) where (N_i) is the data number of the partition i, with $i \in \{1,2,3, \dots, k\}$ and k is the number of partitions performed, and the natural logarithm of the quotient between the recessed range (R_i) and the standard deviation (SD), Ln(R_i/ SD_i), with $i \in \{1,2,3, \dots, k.\}$. The following is the method known as the rescaled range, symbolizedby the expression (R/SD), of the time series described in [13] under consideration X: $\{x_i\}$ is composed byN values. The R/SD statistics follow the power law described in Equation (1), in which R corresponds to the rescaled range, SD is the standard deviation of the data, M is the corresponding partition size, α is a constant and H is the Hurst coefficient [14]; herein, a is a constant and H is the Hurst exponent which represents a fractal measure of long-rangecorrelations in the analyzed released.

$$(R/SD)_M = \alpha M^H \quad (1)$$

3. Results and discussion

The signals of each patient correspond to 12,000 data one minute before and 12,000 data recordedone minute after the intervention; with these data, the coefficient identified as Hurst before getting up (HA) and Hurst after getting up (HD), was calculated in patients with AD. Table 1 shows the HA and HD data for each of the patients operated on.

In Table 1 the coefficient of HA and HD are different, then it shows that there is a change due to the intervention carried out; 50% of the participating patients presented an HA less than 0.5, and in the same way 50% of the patients presented a HA less than HD. The HA average was 0.7429 and the HD average was 0.5060. The ages of the participants range between 70 and 89 years. To demonstrate the reliability of the recorded ECG signal, comparative analyzes of classic indicatorscorresponding to the variability of HR between the rest period and the standing period were carriedout, for this a t-test of measurements was used. repeated and an error probability value of less than 5% was considered for the criterion corresponding to statistical significance ($p < 0.05$), see Table 2.

Table 2, variables such as the average HR and the minimum HR, together with an important trend in the maximum HR, increased during the period of active standing, which represents an autonomic activation as an adaptive response to change; positional suddenness, which is reported in studies that have analyzed vascular and cardiac autonomic functionality in older adults with and without cognitive impairment [12].

Table1. Results of the HA and HD indicators of AD patients.

Patient	Gender	Age	HA	HD
MA	Female	76	0.1702	0.3506
JOS	Female	72	0.2911	0.3693
LE	Female	71	1.1596	1.0192
TRI	Female	70	0.3658	0.5293
YEN	Female	76	1.6216	0.0246
BE	Female	84	0.9594	0.8667
CON	Female	79	0.3765	0.5804
MAR	Female	89	0.9993	0.3083

Table 2. HR feasibility index comparisons.

	Basal Period		Bidepest period			
Variables	Average	SD	Average	SD	Value t	Value p
Average HR	67.941	8.433	73.194	9.934	-4.670	0.001*
HR Maximum	74.266	9.753	79.701	11.613	-2.086	0.067*
HR Minimum	64.522	8.309	68.684	8.830	-5.140	<0.001*
Range HR	9.744	5.176	11.017	5.919	-0.580	0.576
SD HR	2.216	1.312	2.345	1.192	-0.263	0.798

* Statistically significant difference.

4. Conclusions

By using the two methods, fractional Brownian motion and the repeated measures t-test, it is possible to establish in patients with Alzheimer's disease there is a difference in their cardiac activity records with respect to the stimuli and tasks assigned in the study. In particular, the fractional Brownian motion shows a difference now before and after standing, however, since it is not a generalized pattern, it is possible that this indicator cannot contribute to the diagnosis of the disease. These evidenced differences may have exogenous variables not contemplated in the study, which may alter the registration of the signal and therefore the calculation of the H coefficient. With the implementation of the repeated measures t-test, it is determined that with 3 of the 5 strong variables it is possible to have information to contribute to the diagnosis, since the average variables of heart rate, maximum and minimum heart rate presented a statistically significant difference.

The previous study is of great importance because, in general, the training of a neural network is carried out to generate contributions to the diagnosis of the disease from the analysis of magnetic resonance images and patterns of psychophysiological measurements, in which case, the immediate finding is to use the three strong variables for training.

Acknowledgments

The authors offer they're thanks to Universidad Pontificia Bolivariana and the Ministerio de Ciencia, Tecnología e Innovación, Colombia; within the framework of the project "Creación e implementación de un protocolo de evaluación diagnóstica multicomponente en pacientes con trastorno neurocognitivo (TNC) mayor tipo Alzheimer: un estudio multidisciplinar mediante el uso de geometría fractal en imágenes cerebrales".

References

[1] Ministerio de sanidad, política social e igualdad 2012 *Guía de Práctica Clínica sobre la Atención Integrala las Personas con Enfermedad de Alzheimer y otras Demencias* (Cataluña: Ministerio de Ciencia e Innovación)
[2] J Libre 2012 Envejecimiento y demencia: implicaciones para la comunidad científica, la salud pública y la sociedad cubana *Revista Anales de la Academia de Ciencias de Cuba* **2(2)** 1
[3] Gauthier K, Morand A, Dutheil F, Alescio-Lautier B, Boucraut J, Clarys D, Eustache F, Girard N, Guedj E, Mazerolle M, Paccalin M, de la Sayette V, Zaréa A, Huguet P, Michel B, Desgranges B, AGING consortium, Régner I 2019 *BMJ Open* **9(10)** e032265
[4] Toledo M, Junqueira L 2010 *Clinical Autonomic Research: Official Journal of the Clinical Autonomic Research Society* **20(1)** 11
[5] Schwarck S, Busse N, Ziegler G, Glanz W, Becke A and Düzel E 2021 *Front. Sports Act. Living* **3** 684089
[6] Cacioppo J, Tassinary L & Bernston G 2007 *Handbook of Psychophysiology* (New York: Cambridge University Press)
[7] Nonogaki Z, Umegaki H, Makino T, Suzuki Y & Kuzuya M 2017 *Geriatrics & Gerontology International* **17(1)** 92
[8] Simic G, Stanic G, Mladinov M, Jovanov-Milosevic N, Kostovic I, Hof P 2009 *Neuropathology and Applied Neurobiology* **35(6)** 532
[9] Uribe L, Villamizar J, Morantes G, Cerquera A, Prada E, Prada D 2022 *Journal of Physics: Conference Series* **2159(1)** 012010
[10] Giraldi H, Campos E 2015 *Boletín de la Sociedad Mexicana de Física* **29(2)** 103
[11] *F. Yerlikaya Özkurt, C Vardar-Acar, Y Yolcu-Okur, G-W Weber* 2014 *Journal of Computational and Applied Mathematics* **259(B)** 843
[12] Nicolini P, Ciulla M M, Malfatto G, Abbate C, Mari D, Rossi P D, Pettenuzzo E, Magrini F, Consonni D, Lombardi F 2014 *PLoS ONE* **9(5)** e96656
[13] Martin L, Aranda N, Quimbaya C 2015 *Phys. A* **421** 124
[14] Rodriguez E 2012 *Cuadernos Latinoamericanos de Administración* **8(14)** 41

Journal of Physics: Conference Series **2515** (2023) 012009 doi:10.1088/1742-6596/2515/1/012009

Characterization of variables for modeling agroclimatic and phytosanitary events in agricultural crops using deep learning models

T Ramirez-Guerrero[1], M I Hernández-Pérez[1], M S Tabares[1], and E Villanueva[1]
[1] Universidad EAFIT, Medellín, Colombia

E-mail: teramirezg@eafit.edu.co

Abstract. Digital transformation is increasingly present in the agricultural sector thanks to the development and availability of information technologies. However, the agricultural sector is one of the least digitized worldwide, including in countries with highly competitive and technologically advanced agricultural chains and processes. To optimize production using technological tools, it is necessary to combine variables associated with soil, climate, and plant type to estimate crop yields. Therefore, it is important to identify the physical variables that are of interest for the modeling of agroclimatic and phytosanitary events in agricultural crops. This study presents an exploratory type of documentary research to determine what variables are of interest for the modeling of agroclimatic and phytosanitary events in agricultural crops, using a critical analysis based on the results of related studies. Results show the characterization of (i) agroclimatic and biotic variables required to model the mentioned events, (ii) detection of physical evidence associated to biotic factors by means of the analysis of anomalies in the wavelengths of the spectral reflectance of the productive units, (iii) the computational model based on deep learning for the processing of these variables. Furthermore, a data pipeline is proposed that indicates the flow that the characterized variables must go through the analytical tasks.

1. Introduction

Digital transformation is becoming increasingly present in the agricultural sector thanks to the development and availability of information technologies. However, compared to other economic sectors, the agricultural sector is one of the least digitized worldwide, even in countries with highly competitive and technical agricultural chains and processes [1,2]. Given the importance of ensuring food sustainability for the human population, the agricultural sector is more relevant today than in previous years. Under this panorama, the transformation of conventional processes in agriculture by technology-supported techniques can play a fundamental role in increasing crop production in a sustainable manner, while offering economic opportunities to families that depend on this sector [3].

To ensure crop yield, it is important to prevent the occurrence of agroclimatic and phytosanitary events during the management of agricultural crops. These events can lead to affect crop production, affect the marketing of the harvested product, impose restrictions and quarantines on crops and productive zones, and affect the technical-financial analysis of an agricultural crop [4,5]. The main consequence of the last is that insurance and financial products to support the productive unit are very costly and unattractive for the agricultural sector, due to the high operational risk [4]. For this reason, considering that soil variability from one area to another, climate variability and the type of plant grown have an impact on crop yields, it is necessary for farmers to combine these variables and use them to obtain a solution from crop management [3]. This is possible if technological tools are appropriated in the process. By using computational models to process crop variables, it is possible to detect and monitor crop plant diseases, predict the increase of pests and climatic conditions that cause disease and quarantine restrictions, optimize crop yields through predictors, and mitigate the impact of nutrient and input losses before harvest [6-8].

Based on the above, to use a computational model that helps in crop management, it is important to identify what variables are of interest for the modeling of agroclimatic and phytosanitary events in agricultural crops. Therefore, in this study an exploratory documentary research was conducted to characterize the physical variables associated with climatic and biotic factors that allow modeling events in perennial tree crops. The documentary research approached the study of data collection in the crop, the description of the variables associated with physical meteorological phenomena, the detection of anomalies in the plants by means of the wavelengths of the reflectance of the electromagnetic spectrum, and the computational models that can be used for the processing of these variables. Based on the characterization done, a data pipeline is proposed to establish the flow that the variables must follow to obtain crop information that allows recommending appropriate agronomic actions for the crop.

2. Methodology

This study is based on an exploratory type of documentary research, adapted from the methodological proposals followed in the research conducted by [9-12]. Based on these proposals, a qualitative analysis of the data obtained from studies related to the modeling of agroclimatic and phytosanitary events in agricultural crops, the analysis and critical review of research articles, and the analysis and critical review of the results of the studies were conducted [13,14]. Therefore, this study is driven by two main aspects. First, the observation present in the data analysis collected from different academic and scientific sources. Second, the general and systemic view of the theoretical construction provided in the process. Considering the above, the steps defined in the development of this research are described below.

2.1. Documents and data search

Documentary research requires the search for preliminary research on the modeling of agroclimatic and phytosanitary physical variables in perennial agricultural crops. Therefore, in this study, scientific papers were searched in the bibliographic databases. Key words associated with agricultural perennial tree crops, weather variables, pests, diseases, and productivity were used for the search. This search returned a considerable number of results, from which those documents related to the key terms mentioned above were selected in a non-systematic way.

2.2. Data analysis

For data selection and analysis some of the guidelines for the study and meta-analysis of scientific papers, especially in software engineering, have been followed [15,16], in order to perform a correct exploration of the information under commonly used methods. Information obtained was analyzed considering whether it is related to agricultural crops. Particularly, if this information shows whether there is correlation between variables associated with climatic and biotic factors, and the incidence of agroclimatic and phytosanitary events of the target crop. Likewise, the application of analytical models for risk management in agricultural crops was considered.

2.3. Variables characterization

Using the selected and analyzed information, the physical variables associated with climatic and biotic factors that are related to the occurrence of events that may affect productivity in perennial tree crops were characterized. Additionally, through observation field visits in crops of interest -such as Hass avocado- the concurrence of the characterized variables is validated and their impact on the productive unit is estimated. Based on the categorized variables, computational models are studied to help in the prevention of risks in agricultural crops through advanced data analytics.

3. Results

From the data and information collected from the sources mentioned in the previous section, the characteristics of the variables to be used in the modeling of events that may affect agricultural crops were analyzed and interpreted. The agroclimatic and biotic variables associated with physical meteorological phenomena, the sensors to be used for data collection in the field, and the computational modeling were identified. Likewise, observations obtained in the field are reported and a data pipeline is proposed for the automation of the analysis of the physical variables of interest.

3.1. Agroclimatic and biotic variables

Based on the data reported in [7,17-20], the information corresponding to the agroclimatic and biotic variables that affect productivity in agricultural crops was selected. These variables can be studied through analytical models to mitigate negative impacts on crops and improve yields.

The characteristics of physical meteorological phenomena that are involved among the causes of agroclimatic events in crops correspond to weather characteristics and soil characteristics [18]. These characteristics, together with crop management characteristics, are useful for yield estimation. Therefore, within the weather data, the variables of interest include maximum and minimum temperature, solar radiation, rainfall, and relative soil moisture [18].

From the variables mentioned above, soil moisture is one of the critical variables in crops of high commercial interest, such as Hass avocado. Excess water produced by the accumulation of excess rainfall or irrigation that causes over-wetting of croplands with gentle to moderate slopes constitutes one of the problems that most affect the decrease in yields in tropical and subtropical areas [17]. Most species not adapted to high soil moisture conditions are at elevated risk of survival because excess water deprives the roots of oxygen needed for respiration, leads to nitrate depletion and the arrival of pathogens leading to reduced crop production [17].

To study precisely how these climatic variables impact performance in a year, the values of each variable must be collected in a year during fifty-two weeks and accounted for in features set [18]. With these variables, together with the soil characteristics, it is possible to obtain diverse applications that support crop improvement. For example, in [19,21], the predictions of the proposed model were able to identify areas where Hass avocado is grown commercially, highlighting that this model was strongly influenced by the temperature and precipitation variables, with a greater predominance of the first variable. This is consistent with previous understanding of the distribution and ecology determinants of this type of crop [21].

Moreover, among the biotic factors, the presence of pests and pathogens are the main cause affecting productivity in agricultural crops. In the case of Hass avocado, the evaluation of the influence on the yield of production units with respect to the distribution and abundance of the main pests and parasitoids of this avocado variety was studied in [20]. Based on the proposed hypothesis, the relationship between the normalized difference vegetation index and temperature and rainfall is presented, and how crop productivity varies according to avocado pest populations. Among the findings reported, it was found that the average mass per fruit varied between 150-190 grams, in the production units where the presence of fruit flies and parasitoids found during fruit hatching was recorded, where only 7% of the pupae did not hatch [20].

Another biotic factor that affects production in agricultural crops are weeds that tend to grow and spread invasively over large areas of land. These weeds grow rapidly and compete with the crop plants for environmental resources such as soil nutrients, water availability, sunlight, and space [7]. Weed control is important to avoid yield reduction in crops, either by mechanical treatment or by the application of agrochemicals, the latter being the most commonly used operation [7].

3.2. Data collection sensors

Identifying phytosanitary events (such as diseases) in a timely manner is a key point for the effective management of a crop. These events can be caused by several types of pests, fungi, viruses, and other biotic factors. Physical evidence of disease, resulting from pathogens presence and changes in plant phenotype, can be leaf and fruit spots, wilt, and canopy color variation [7]. For the early detection of the presence of these risk causes, both remote sensors and in situ sensors can be used to prevent a sanitary event by monitoring agroclimatic and biotic variables.

Remote sensing agricultural data collection techniques, employing satellites and unmanned aerial vehicles (UAV) combined with weather data, generate availability of real-time files and datasets that together with digitized soil data and other data streams from in situ sensors, can provide an improved understanding of crop-weather-soil interactions quickly and cheaply [3]. Thus, the use of UAV together with multispectral cameras makes it possible to detect anomalies in the productive units using spectral reflectance. As can be seen in avocado, where differences between healthy and diseased laurel wilt plants can be found in some 10 nm wavelengths of the red-edge (740 nm, 750 nm) and near-infrared (760 nm, 940 nm, 950 nm) bands [22]. Together with 4-class classification algorithms, zone-specific control tactics can be applied depending on disease development. Furthermore, employing electronic sensors for collection and wireless transmission of crop physical variables data associated with weather factors, such as soil moisture and temperature, allows more inputs to be available in the model to identify disease causes [21] and recommend appropriate agronomic practices for risk management.

3.3. Computational modeling

For processing and analyzing data provided by agroclimatic and biotic variables in an agricultural crop, artificial intelligence computational models are key tools in the process. Machine learning (ML) and deep learning (DL) models are capable of processing data from remote sensors and in-field sensors, supporting transfer learning tasks aimed at improving human capabilities in the diagnosis and detection of phytosanitary and agroclimatic events in agricultural crops [23]. It is possible to do this using ML/DL algorithms that take crop data as input and make predictions based on these data using forecast maps as a reference. Equation (1) presented in [23] defines the general structure for a stacked deep learning computational model, where: yr_k defines the softmax function for units classification in the crop by damage categories; $x_{j_0,k}$ represents the control points in a vegetation index for a time k; C_{j_m,j_n}, w_{j_1,j_0} represent the supports of the structure for the model, where w are the weights for the internal layers and C are the weights for the output layer.

$$yr_k = \text{softmax}\left(\sum_{j=1}^{NO} C_{j,k}\left(\sum_{j_3=1}^{NO_3} w_{j_3,j_2}\left(\sum_{j_2=1}^{NO_2} w_{j_2,j_1}\left(\sum_{j_1=1}^{NO_1} w_{j_1,j_0}\cdot x_{j_0,k}\right)\right)\right)\right). \quad (1)$$

Likewise, the internal structure of the stacked deep learning model can be composed by known models of artificial neural networks, where each model in particular is responsible for the processing of a crop variable, and the results are integrated with the passage through the following layers of the model. Convolutional neural networks (CNN), defined in Equation (2), are specially designed for classification, detection and segmentation of elements using computer vision techniques. In turn, recurrent neural networks, defined in Equation (3), are specially designed to manage sequential data, where the previous step output serves as input to the current step [24,25].

$$y_i = X * W_i + b_i,\ \ i = 1, \dots, m, \quad (2)$$

$$h_t = \tanh(W_x x_t + W_h h_{t-1} + b_t). \quad (3)$$

3.4. Field visit observations and data pipeline proposal

Through observations during the field visit to a Hass avocado crop, the impact of biotic factors (pests, diseases) and agroclimatic factors (excess moisture, high solar radiation) on some trees of the production unit was evidenced. The observations validated the relevance of the agroclimatic, and biotic variables characterized in this study, which can be used to predict events that decrease yields. Figure 1 shows the damage caused by three agents on some trees of the crop. First image corresponds to the presence of beetles that consume the leaves and peel of the fruit, causing defoliation of the plant and deformation of the fruit, impeding its commercialization. Second image corresponds to the wilt of a tree caused by excess humidity, causing the loss of the tree canopy and exposure to high solar radiation. Third image, taken with the use of a 40x entomological lens, shows the presence of microscopic pests (black spots) on the leaves, causing deformation of the fruit during growth and changes in the phenotype of the tree.

To take advantage of the information coming from the characterized variables, it is necessary to establish the data flow through the data analysis tasks. For this purpose, it is possible to employ a data pipeline that directs the step-by-step flow of crop data of interest, from its collection to its use in recommending appropriate agronomic practices. Figure 2 shows a proposal for the data pipeline, where the first step corresponds to the capture of crop data by remote sensing and in-field sensors. Second step indicates the tools for storing the collected data. Third step corresponds to the data cleaning and the training of the appropriate computational model for each type of crop. Fourth step establishes the statistical metrics used to trained model evaluation, as well as the prediction percentage when new values of the characterized variables are entered. Fifth step refers to the application of the model for crop management, using the most appropriate technological tool.

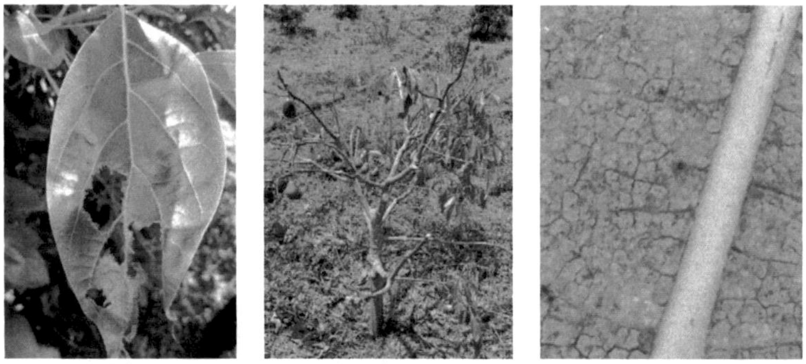

(a) (b) (c)

Figure 1. Damage caused by phytosanitary and agroclimatic events in an avocado crop; (a) damage caused by beetles (Coleoptera: Melolonthidae) on leaves, (b) wilt caused by water stress and high solar radiation, and (c) presence of microscopic pests (Thysanoptera: Thripidae) on leaves.

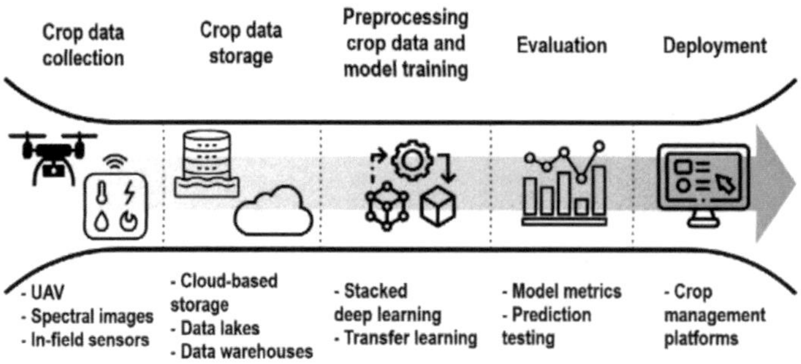

Figure 2. Proposed data pipeline for the analysis of agroclimatic and biotic variables in agricultural crops.

4. Conclusions

This study characterized the variables associated with agroclimatic and biotic factors that can be used in the modeling of events caused by physical meteorological phenomena in agricultural crops. The characterization of physical variables of interest (excess soil moisture, solar radiation, rainfall, presence of pests, among others) is the first step to define the type of processing of data captured in the field using deep learning computational models.

Thus, identifying the tools to be used to capture data associated with these physical variables requires extensive support from current advances related to remote sensors in unmanned aerial vehicles, which in conjunction with multispectral cameras make it possible to collect data on the physical processes of the crops, and from these to monitor anomalies in the wavelengths according to spectral reflectance in the crop units studied. Once the physical variables and the tools to collect data on these variables in the crop have been defined, it is possible to establish a computational model based on deep learning to process the data and obtain predictions that allow taking appropriate agronomic actions to prevent negative impacts on crop productivity due to an agroclimatic or phytosanitary event correlated with the physical variables of interest.

For the next stages of this research, it is suggested to employ additional advanced analytical techniques specific to the processing of wavelength variations obtained by multispectral and hyperspectral reflectance measurement instruments, as well as techniques that address the processing of physical data obtained in the field and in external open access sources. This in order to check if the use of the characterized variables associated with physical phenomena is feasible for both the established computational model and the proposed data pipeline for data analysis in agricultural crops. This would mean an advance in the study of physical phenomena for the improvement of agronomic processes.

Acknowledgments

Universidad EAFIT funds this study. The authors extend thanks to the Vicerrectoría de Ciencia, Tecnología e Innovación of Universidad EAFIT for funding the project and providing the PhD scholarship to the first author. Additionally, the authors extend thanks to PhD Alejandro Peña-Palacio, PhD Alejandro Marulanda, and PhD Mauricio Toro, for their help and ideas in this project.

References

[1] Ayaz M, Ammad-Uddin M, Sharif Z, Mansour A, Aggoune E-H M 2019 Internet-of-things (IoT)-based smart agriculture: toward making the fields talk *IEEE Access* **7** 129551
[2] Cravero A, Sepúlveda S 2021 Use and adaptations of machine learning in big data-applications in real cases in agriculture *Electronics* **10(5)** 552
[3] Cheema M J M, Khan M A 2019 Information technology for sustainable agriculture *Innovations in Sustainable Agriculture* ed M Farooq, M Pisante (Cham: Springer) p 585
[4] Pena A, Tejada J C, Gonzalez-Ruiz J D, Gongora M 2022 Deep learning to improve the sustainability of agricultural crops affected by phytosanitary events: a financial-risk approach *Sustainability* **14(11)** 6668
[5] Van Klinken R D, Fiedler K, Kingham L, Collins K, Barbour D 2020 A risk framework for using systems approaches to manage horticultural biosecurity risks for market access *Crop Protection* **129** 104994
[6] Shrestha G, Deepsikha, Das M, Dey N 2020 Plant disease detection using CNN *Applied Signal Processing Conference* ed Dey D, Dalai S, Ray S, Chatterjee B (Kolkata: Institute of Electrical and Electronics Engineers) p 109
[7] Benos L, Tagarakis A C, Dolias G, Berruto R, Kateris D, Bochtis D 2021 Machine learning in agriculture: a comprehensive updated review *Sensors* **21(11)** 3758
[8] Karnati R, Rao H J, Prakash P G O, Maram B 2022 Deep computation model to the estimation of sulphur dioxide for plant health monitoring in IoT *International Journal of Intelligent Systems* **37(1)** 944
[9] Ahmed J U 2010 Documentary research method: new dimensions *Indus Journal of Management & Social Sciences* **4(1)** 1
[10] Gómez L 2011 Un espacio para la investigación documental *Vanguardia Psicológica* **1(2)** 226
[11] Dueñas Ramírez L M, Villegas López G A 2020 Computer science development and technologies associated with industry 4.0 applied to industrial maintenance in Colombia *Journal of Physics: Conference Series* **1513(1)** 012002:1

[12] Ramirez-Guerrero T, Toro M, Villegas López G A, Castañeda L F 2020 Low-cost computational systems applied to physical architectures in public transportation systems of intermediate cities *Journal of Physics: Conference Series* **1702(1)** 012018:1
[13] Corbin J, Strauss A 2014 *Basics of Qualitative Research: Techniques and Procedures for Developing Grounded Theory* (Thousand Oaks: SAGE Publications)
[14] Garcés Cano J E, Duque Oliva E J 2007 Metodología para el análisis y la revisión crítica de artículos de investigación *Innovar* **17(29)** 184
[15] Moher D, Shamseer L, Clarke M, Ghersi D, Liberati A, Petticrew M, Shekelle P, Stewart L A 2015 Preferred reporting items for systematic review and meta-analysis protocols (PRISMA-P) 2015 statement *Systematic Reviews* **4(1)** 1
[16] Kitchenham B, Brereton O P, Budgen D, Turner M, Bailey J, Linkman S 2009 Systematic literature reviews in software engineering-A systematic literature review *Information and Software Technology* **51(1)** 7
[17] Herrera J, Duarte C, González F, Cid G 2016 Efecto del exceso de humedad del suelo sobre el rendimiento en algunos cultivos de importancia agrícola en Cuba *Revista de Ingeniería Agrícola* **6(2)** 3
[18] Manoj M, Makkithaya K, Narendra V G 2022 A federated learning-based crop yield prediction for agricultural production risk management *Delhi Section Conference* (New Delhi: Institute of Electrical and Electronics Engineers) p 1
[19] Ramírez-Gil J G, Cobos M E, Jiménez-García D, Morales-Osorio J G, Peterson A T 2019 Current and potential future distributions of Hass avocados in the face of climate change across the Americas *Crop and Pasture Science* **70(8)** 694
[20] Toukem N K, Yusuf A A, Dubois T, Abdel-Rahman E M, Adan M S, Mohamed S A 2020 Landscape vegetation productivity influences population dynamics of key pests in small avocado farms in Kenya *Insects* **11(7)** 424
[21] Ramírez-Gil J G, Giraldo Martínez G O, Morales Osorio J G 2018 Design of electronic devices for monitoring climatic variables and development of an early warning system for the avocado wilt complex disease *Computers and Electronics in Agriculture* **153** 134
[22] De Castro A I, Ehsani R, Ploetz R, Crane J H, Abdulridha J 2015 Optimum spectral and geometric parameters for early detection of laurel wilt disease in avocado *Remote Sensing of Environment* **171** 33
[23] Peña-Palacio A 2021 Criterios para la configuración de plataformas de inteligencia aumentada para el mejoramiento de la sostenibilidad de cultivos agrícolas *Congreso Internacional de Ingeniería de Sistemas* (Lima: Universidad de Lima) p 25
[24] Alibabaei K, Gaspar P D, Lima T M, Campos R M, Girão I, Monteiro J, Lopes C M 2022 A review of the challenges of using deep learning algorithms to support decision-making in agricultural activities *Remote Sensing* **14(3)** 638
[25] Wang D, Cao W, Zhang F, Li Z, Xu S, Wu X 2022 A review of deep learning in multiscale agricultural sensing *Remote Sensing* **14(3)** 559

Journal of Physics: Conference Series **2515** (2023) 012010 doi:10.1088/1742-6596/2515/1/012010

Computational mathematics applied in the identification of the risk of indirect contagion of COVID-19 in university classrooms

S Arceo-Díaz[1], E E Bricio-Barrios[1], J Carrillo-Alcaráz[1], O Lua-Madrigal[1], and J R Moreno-Peña[1]
[1] Tecnológico Nacional de México, Colima, México

E-mail: olimpo.lua@colima.tecnm.mx, elena.bricio@colima.tecnm.mx

Abstract. This exploratory study evaluated the risk of contagion from airborne diseases, such as coronaviruses, in schools. For three days, the concentration of carbon dioxide in two university classrooms was monitored for 90 minutes, while the students took their math classes. We use these values to validate a first-order model for carbon dioxide concentration and calculate the air exchange rate indirectly (avoiding the need for expensive measurement equipment). The air exchange rate obtained allowed us to assess whether the usual ventilation systems (both natural and mechanical) are sufficient to guarantee a low risk of contagion of aerosols due to respiration. The results show that the risk of contagion is low if three factors are considered: the level of conversation within the classroom, the usage of a moisture extraction system, and the lecture duration. The risk is low if the lecture time is less than 50 minutes, the level of conversation is moderate, and a moisture extraction system is available. If these conditions are not met the risk is considerably higher even if mechanical ventilation is employed.

1. Introduction

The study of transport phenomena is one branch of physics that allows optimizing the costs associated with any complex process by estimating the relevant variables to calculate predictive models [1]. In specific, epidemiology has used some mathematical models, based on the properties of the fluids in closed spaces, to describe the speed of transference of air-borne illnesses [2].

The concentration of carbon dioxide (CO_2) can be used as an indicator of ventilation. In an enclosed space, the concentration of this gas increases with the exhalation rate of the occupants and, in the absence of other sources, decreases rapidly if they are removed and ventilation is adequate [3]. Likewise, this rate of decrease depends directly on the speed with which the total air volume in the room can be replaced by the action of ventilation, whether natural or mechanical. This replacement is called the air change rate per hour (ACH) and can be used, in conjunction with a mathematical model, to establish a relationship between the concentration of CO_2 and ventilation [4].

In addition to ventilation, the concentration of CO_2 in an indoor space has been associated with the risk of contagion of the severe acute respiratory syndrome, known as SARS-CoV-2 [3]; although this work did not identify a model that directly relates these two variables, the World

Health Organization (2020) reported that a concentration of 600 0 parts per million (ppm) represents 0.5% of the air breathed by another individual, a percentage that doubles and triples at 800 ppm and 1000 ppm. Whereas, values higher than 2000 ppm are considered a risky means of contagion of respiratory diseases [5].

Keeping the CO_2 as low as possible allows for optimizing the protection provided by ventilation since the transmission of the virus is more likely inside buildings. This is why practical, affordable, and widely applicable measures are urgently needed to monitor and limit the risks of indoor transmission. Previous studies [6] have suggested the use of air purifiers (which use a fan, an air pump, a water pump, and an ultrasonic generator) to control the concentration of CO_2 [7]. These prototypes have advantages such as low manufacturing cost or easy installation but their reported performance in reducing the concentration of CO_2 is limited.

The present work presents the results of an exploratory investigation that used the measurements of the CO_2 concentration within two university classrooms to propose a model for estimating the inflow and outflow of available mechanical ventilation devices. The design equations of the saturation rate of this gas were incorporated, taking into account parameters such as the number of people, age range as well as the dimensions of the classroom, under three different scenarios. These results will be used in the first step of a project funded by the National Technology of México to set a sensor network for monitoring the CO_2 concentration and estimating the risk of contagion of air-borne diseases.

2. Methodology

Two classrooms within the Technological Institute of Colima were used as a workbench. The experimental tests to measure the concentration of CO_2 were carried out during the last three days of the annual course of admission to the institution. Structurally, each classroom has an area of $7.43 \times 8.35\ m^2$ and an approximate volume of 155 m^3. It offers a natural ventilation system of drafts from south to north made up of two pairs of windows of 1.6 m^2 [8].

The distribution of the natural and mechanical ventilation systems as well as the location of the measuring system is shown in Figure 1, where "X" and "W" mark the locations of the CO_2 sensor and the windows, the arrows indicate the entry and exit of mechanical ventilation generated by two air-conditioned systems and two fans anchored to the ceiling. Additionally, the air conditioning devices had the function of absorbing the environmental humidity level; to test if the usage of the ceiling fans caused any significant difference in the ventilation, it was planned that they were permanently on for group A and off for group B.

Two groups of 35 students were selected to measure the CO_2 concentration. These groups (labeled as "A" and "B"), remained in identical classrooms for a period of about one hour and 40 minutes. Three different scenarios were prepared to measure the rate of generation of CO_2 inside a classroom: One with an intermediate recess period, in which students can leave the classroom (labeled as "1"), one in which half of the time students remained in silence ("2") and one in which students remained in silence for the complete duration of the session ("3").

The rate of generation of CO_2 was measured by using the Nobrand brand meter [9]. This device was installed on the teacher's table located on one of the edges of the classroom and the strategy of recording the data of interest was carried out in two stages: the identification of the concentration conditions of CO_2 initial, that is, without students and the saturation rate of this gas during the session with the students inside.

Since this work proposes to estimate the rate of replacement and actions to minimize the concentration of CO_2 inside the classroom, it is necessary to identify the airflow generated by the mechanical ventilation device. However, due to the wear caused by continuous usage, it would be necessary to verify their fabric values by employing a balometer, whose cost is approximately 5,000 USD.

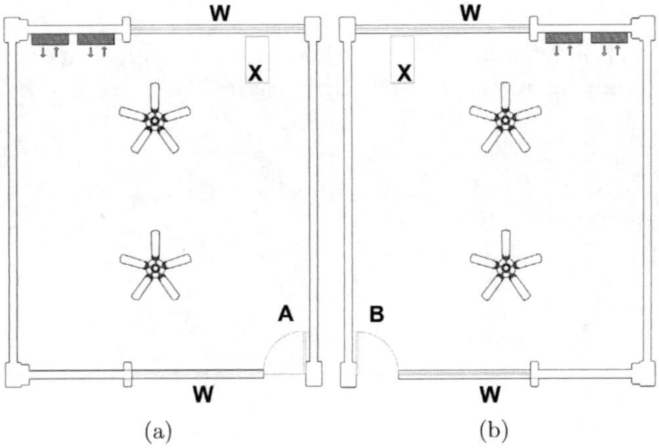

Figure 1. Diagrams of the two university classrooms and distribution of (a) natural and (b) mechanical ventilation systems.

We solved this problem through the indirect estimation of the airflow entering and leaving the classroom (Q_{in} and Q_{out}, respectively), measured in $L \cdot min^{-1}$. We used the first-order mathematical model proposed by Suebyat in [10]. This model proposes that the total concentration depends on the CO_2 in the air within the classroom (C_a), the CO_2 generated by the respiration of its occupants (C), and that being propagated by the air flow entering from the windows (C_e). Equation (1) shows the differential equation that relates these variables.

$$V\frac{dC}{dt} = n * p * C_a + Q_{in} * C_e - Q_{out} * C, \tag{1}$$

where V is the volume of the classroom in m^3, n is the number of people and p is the breathing speed in $L \cdot min^{-1}$. On the other hand, if the concentration of CO_2 inside the classroom is homogeneous, it can be calculated from the Equation (2) proposed by Issarow in [11]:

$$C_a = \frac{Q_{in} * C - C_e}{n * p}. \tag{2}$$

By substituting Equation (2) in Equation (1), we obtained a function that does not depend on the number of occupants (Equation (3)), nor on the respiration rate (parameters that in this case study were constant during the measurements).

$$\frac{dC}{dt} = \frac{1}{V}\left[(Q_{in} * C - C_e) + Q_{in} * C_e - Q_{out} * C\right]. \tag{3}$$

By having the kinetics of CO_2 under the aforementioned experiments, these trajectories will make it possible to indirectly estimate Q_{in} and Q_{out} per least squares by means of the lsqcurvefit function, available in Matlab [12]. Finally, the replacement rate of the classroom was calculated, which from its magnitude, dimensionless, is established with a scale proposed by Allen in [4] from low ($ACH \geq 3$) to ideal ($ACH = 6$) by means of Equation (4).

$$ACH = \frac{(Q_{out} - Q_{in}) * 60 * Q_{in}}{V * Q_{out}}. \tag{4}$$

3. Results

Figure 2 shows the temporal variation of the CO_2 concentration; the left panel corresponds to the three sessions of group A, while the one on the right corresponds to group B.

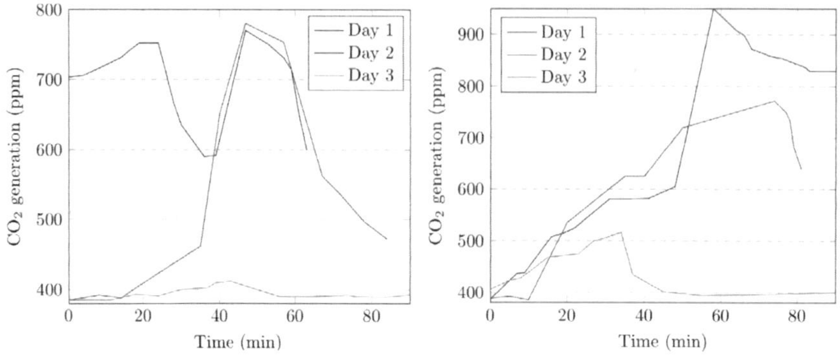

(a) (b)

Figure 2. Temporal variation of the concentration of CO_2 in classrooms (a) A and (b) B.

The students of group A were able to enter the classroom about half an hour before the start of the first session; the level of conversation of the students during this period caused the CO_2 concentration to start near 700 ppm. The CO_2 continued increasing during the first 25 minutes, causing the teacher to allow a 15-minute recess. The CO_2 dropped from 750 ppm to 590 ppm during this recess. After resuming activities, the CO_2 concentration rose again to 770 ppm. During the last ten minutes, the CO_2 dropped under 600 ppm after the teacher opened the windows and the students left the classroom. For group B, the CO_2 concentration peaked at 620 ppm during the first hour. After a failure in one of the two air conditioning devices, the CO_2 rose to 950 ppm. This caused the teacher to start the 15-minute recess period. However, unlike group A, the decrease in the CO_2 concentration was less than 100 ppm. As activities resumed and mechanical ventilation was reactivated, CO_2 maintained its downward trend, albeit with a shallower tendency than during the recess; as with group A, the CO_2 concentration dropped rapidly in the 10-minute departure period.

The second session for both groups started with a 40-minute quiz; the teacher asked the students of group A to remain silent while allowing those in group B to have a moderate level of conversation while answering the exercises individually. In both groups, the CO_2 concentration remained at values below 600 ppm. During the second 40-minute period, the teacher asked the students of both groups to join in teams to solve a second set of problems. In both groups, the CO_2 concentration rose continuously until reaching values close to 750 ppm at the end of the session. For both groups, the CO_2 concentration dropped rapidly during the 10-minute leave period.

In the third session, which consisted of the course's final evaluation, the two groups show similar results. As speaking was not allowed in both cases, the concentration of CO_2 rose slowly, peaking at 410 ppm, for the first group, and 510 ppm, for the second, in less than 40 minutes and decreasing to the environmental level as students left the classroom.

The box plots in Figure 3 show the distribution in the values of the concentration of CO_2 in each of the classrooms during the three days in which the experiment was performed; the horizontal and vertical axes correspond to the measurement day and the CO_2 concentration, respectively. The box plots show the distribution of the data for both groups concerning the median; in all cases, the distribution was asymmetric and no general trend can be observed that repeats itself, either by day or group. The boxplots show that no significant difference can be observed in the measurements for the CO_2 concentration inside each classroom during the first day (706 ppm for group A against 702 ppm for group B); for the rest of the days, the median CO_2 concentration for group B is significantly higher than that of group A.

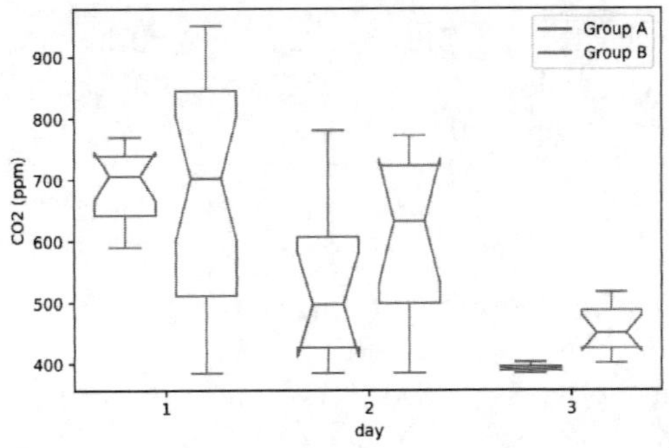

Figure 3. Box plots of the distribution of recorded CO_2 concentration.

To test if there was a statistically significant difference in the measurements coming from each group, the Mann-Withney and t-student tests were used [13]; the results are in Table 1. The working hypothesis, that lectures from group A are significantly lower than those from group B, was tested; only the measurements from the third sessión showed strong enough evidence for this.

Table 1. U Mann-Withney and t-student hypothesis test of the case studies.

Test	Day 1	Day 2	Day 3
t-Student	0.61	0.25	4×10^{-4}
U-test	0.51	0.19	2×10^{-5}

By having the volume of the classroom, the number of people inside, and the measurements of CO_2 concentration, the fitting of the experimental data and the mathematical model described in Equation (1) to Equation (3); Table 2, shows the flow of entry, exit and replacement rate in six activities that are carried out during the school period.

Table 2 shows that, in all cases, the flow of entry and exit differ from each other, this indicates that the capacity that mechanical ventilation (Q_{in}) acts independently of the number of people inside and, depending on the humidity inside, the extraction system will be activated. On the other hand, classroom B where the ceiling fans were activated only fulfills the function of keeping the air in circulation and favors the outflow (Q_{out}), but this is insufficient if the rate of air renewal is taken into account. In this regard, the ACH identified that only on the second-day classroom B during the evaluation reached the highest value, indicating that, when reaching 700 ppm of CO_2 the session can be suspended for 10 min and return to activities safely. On the other hand, the rest of the tests do not reach the minimum value of 3 units requiring students to stay out of the classroom before the academic session.

Table 2. Estimation of flows generated by natural and mechanical ventilation during each session.

Parameter	A			B		
	Day 1	Day 2	Day 3	Day 1	Day 2	Day 3
Q_{in}	251	212	224	253	251	164
Q_{out}	154	234	247	164	94	81
ACH	1.60	0.51	0.27	1.76	2.74	1.10

4. Conclusions

This work proposes the use of carbon dioxide measurements to indirectly estimate the risk of COVID-19 contagion inside a classroom; based on the record of data obtained from two groups of students during three days of classes, no statistically significant difference in carbon dioxide concentration was found, despite the time difference (morning and summer) or the lack of use of the ceiling fans in one of the classrooms. However, the temporal evolution of the carbon dioxide concentration suggests that the risk of contagion from air-borne diseases can be reduced if the duration of lecture time is under 50 minutes, the degree of speaking is moderate and the mechanical ventilation devices are equipped with an absorption system for environmental humidity. Whereas, when evaluating the rate of air renewal per hour, it is recommended to reduce the volume of students to achieve adequate air quality conditions inside the classroom or to incorporate an air extraction system that increases the rate of air renewal.

As future work, we will evaluate the temporal variation of concentration of carbon dioxide making variations in the number of occupants, volume of the classroom, and use of different mask materials. While for the continuous monitoring of the carbon dioxide concentration, a network of sensors will be created which, along with the proposed model, will make it possible to calculate the CO_2 transfer gradient and identify the areas where the potential risk of contagion is greater. Likewise, this network will provide measurements of both temperature and humidity, which will help establish measures to ensure the thermal comfort of students within the classroom.

Acknowledgments

The authors wish to thank the financial support generated by the National Technology of México through project 15414.22-P: "Prototipo de sistema automatizado para controlar la calidad del aire en un salón de clases a través del monitoreo y extracción del CO_2, como estrategia de prevención ante el contagio por COVID-19." and the resources and installations provided by the Technological Institute of Colima and its Departament of Mechatronics.

References

[1] Bird R 2002 *Fenómenos de Transporte* (México: Reverte)
[2] Feng Z Castillo C, Capurro A 2000 A model for tuberculosis with exogenous reinfection *Theoretical Population Biology* **57** 235
[3] Peng Z, Jiménez J 2021 Exhaled CO_2 as a COVID-19 infection risk proxy for different indoor environments and activities *Environmental Science and Technology Letters* **8** 392
[4] Allen J Spengler J Jones E, Cedeno J 2020 *Guía en 5 Pasos para Medir la Tasa de Renovación de Aire en Aulas* (Massachusetts: Harvard University)
[5] Zhao M Wang M Zhang J Ye J Xu Y, Wang Z 2020 Advances in the relationship between coronavirus infection and cardiovascular diseases *Biomedicine and Pharmacotherapy* **127** 110230
[6] Jumlongkul A 2021 *Low-cost Air Purifier Prototype Using a Ventilating Fan and Pump Against COVID-19 and Haze Pollution School of Medicine* (Thailand: Luang Universit)
[7] Rony R Yang H Krishnan S, Song J 2019 Recent advances in transcritical CO_2 (R744) heat pump system: a review *Energies* **12** 457

[8] Ochoa J 2014 *Planos Arquitectónicos Escala 1:1000 del Instituto Tecnológico de Colima* (México: Instituto Tecnológico de Colima)
[9] Dongguan Jinlide Electronic Technology 2022 *Carbon Dioxide Detector* (China: Dongguan Jinlide Electronic Technology)
[10] Suebyat K Oyjinda P Konglok S, Pochai N 2020 A mathematical model for the risk analysis of airborne infectious disease in an outpatient room with personal classification factor *Engineering Letters* **28** 1331
[11] Issarow C Mulder N, Wood R 2015 A mathematical model for the risk analysis of airborne infectious disease in an outpatient room with personal classification factor *Journal of Theoretical Biology* **372** 100
[12] Pérez C 2002 *Matlab y sus Aplicaciones en las Ciencias e Ingeniería* (Madrid: Pearson Educación)
[13] Winters R Winters A, Amedee R 2010 Statistics: a brief overview *Fall* **10** 213

Variational study of the pendulum's angular motion in a differential equations course

R Fallas-Soto[1,2], and C Gaete-Peralta[3]
[1] Departamento de Educación Matemática, Universidad de Costa Rica, Montes de Oca, Costa Rica
[2] Centro de Investigaciones Matemáticas y Metamatemáticas, Universidad de Costa Rica, Montes de Oca, Costa Rica
[3] Departamento de Matemáticas y Ciencias de la Ingeniería, Universidad Bernardo O' Higgins, Santiago, Chile

E-mail: rodolfo.fallas@ucr.ac.cr, claudio.gaete@ubo.cl

Abstract. The objective of this research is to get students to mean the mathematical objects related to the second order differential equation through the context of a physical phenomenon. With the support of the physical system of the simple pendulum, the behavior of the angular position is studied in a modelling process with the help of simulators. From variational thought and language as a line of research of socio-epistemology in mathematics education, activities that promote a modelling process in a differential equations course for students in initial training for teaching mathematics are designed, implemented, and analyzed. The results of the process experienced by the students are presented, such as: familiarization of the phenomenon, use and interpretation of the simulations, construction of the mathematical model and its predictive role for decision making. It was identified that the students' experiences, as well as the physical context, were determining elements for success in the resolution of the design.

1. Introduction

The teaching of differential equations traditionally focuses on the repetition and diversity of algebraic methods to calculate their general or particular solution (given an initial value). Algebraic solutions of differential equations can be found even without a deep understanding and conceptualization of the equations, so students do not feel the need to understand the equation and related concepts [1-3]. In the same way, the applications of differential equations have an artificial use, that is, they lack a real situational context, where the student can make comparisons of theoretical models with the real or physical model. The mathematical tasks used, generally offer the differential equation and the initial values associated with the physical model, focusing on the calculation of a particular solution.

The construction or understanding of the construction of the differential equation and the interpretation or use of its solution, associated with a physical phenomenon, must be carried out for the teaching, and learning of differential equations. Perceiving movement and studying the change in a variable play an important role in the use that the mathematical model can have for predictive purposes [4]; having a better understanding of the role of real-world conditions and the behavior of the phenomenon helps to make the appropriate assumptions in the analytical method of undetermined coefficients [5]. The mathematical tasks that consider the above, help students to use the differential equations to make decisions and obtain short-term predictions about the physical phenomenon [6], for example. This organization of tasks is part of mathematical modelling, where students conjecture, interpret, control, validate and confront models (real and physical) [7].

The aim of the work is to get students to mean the mathematical objects related to the second order differential equation through the angular position of the pendulum from a mathematical modelling task to their initial formation of mathematics teachers at Universidad de Costa Rica, in the virtual mode of the course. For this, the five axes of formation of the mathematical education career must be considered, which corresponds to the vision that the university teacher must consider for the design, development and evaluation of the courses taught in this career [8]: didactic-mathematics, information and communication technologies, application of mathematics, history-epistemology of mathematics and teaching occupation.

This work is framed within the discipline of mathematics education, using the line of variational thinking and language from socio-epistemological theory in mathematics education (STME). This theory is based on an epistemology of practices to explain the social construction of mathematical knowledge. STME analyses the practices that accompany objects, which allowed us to move from examining the apprehension of the object itself (school mathematical knowledge) to an in-depth analysis of the social use of that object (sociocultural mathematical knowledge) [9]. Variational Thinking and Language born within STME arise when carrying out a process that leads to the transition from an E_i state to an E_j state and vice versa. The prediction of the evolution of complex change systems is sought, as an elemental necessity of operation, of a focus on the way to vary even over the same variable [10].

In terms of this work, the following questions are considered for the design of learning situations that start from the qualitative study of change to its quantitative study: what changes? How it changes? How much does it change? In addition, for the study of these situations it is important to consider the construction of the reference system: declaration of variables and constants, the relationship between variables, identification and use of the variation orders, use of measurement units and reference points. On the other hand, STME studies mathematical modelling, focusing on the arguments of students to support their decisions, that is, studying the nuances that mathematical concepts acquire when they are used in social and cultural contexts, in which a form mathematical knowledge is implicit [11].

2. Methodology

The activities were developed in the ordinary differential equations course (code MA-0030) corresponding to the mathematical education career at the University of Costa Rica. The classes were taught virtually from April to July 2020 and were attended by 6 students (3 men and 3 women). The study of the angular positioning of the simple pendulum was developed by the 3 students as part of the extra class work, while the 3 students studied the mass-spring simple harmonic motion. Before developing this activity, in previous weeks the applications were studied: population growth, logistic curve, chemical mixtures, cooling law, electric series circuits resistor-capacitor (RC) and the resistor-inductor-capacitor (RLC); during the study of these applications, five moments were developed, based on the real situational context and context of significance in the study of change [12].

- Sensitization to the study phenomenon: with the support of interactive scientific simulations PHET [13] created (free apps) by the University of Colorado Boulder, the students were able to perform virtual experiments in which they were able to perceive movement and provide information about the behavior of the phenomenon.
- Change qualification: at this moment, the students had to declare the constants and variables in play, in addition to the different orders of variation in a qualitative way: it grows more, it grows less, it decreases more, it decreases less. Units of measurement for variables and constants are identified, as well as reference points such as the role of initial values.
- Change quantification: with the support of the Vensim digital simulator (free personal learning edition), which uses a block-oriented language, students are provided with graphical modelling with stock, flow and causal loop diagrams that simulate the behavior of the variable in study. This

should be explored by the students and modified, if necessary. This simulator is based on a declarative programming language.

- Construction of the differential equation: with the support of physical results and previous moments, each of the procedures to build the differential equation that models the phenomenon under study is studied.
- Decision making: the differential equation is solved, and the solution is interpreted with the graphs obtained by the Vensim simulator in the third moment. The role played by the initial values is reflected and predictions about the behavior of the phenomenon are given. In addition, they redesign examples of the application in the study of textbooks, so that they have a more functional character in terms of context.

The evaluation instrument developed to qualify and analyze the work developed by the students was as follows in Table 1. In each indicator, 0 points would be obtained in case of not responding to what was requested.

Table 1. Evaluation rubric for the modelling activity.

Indicator	3 points	2 points	1 point
Knowing the phenomenon	Declare the variables, constants, units of measure, variation orders, and initial value. It offers a first graph of what the behavior of the phenomenon would be like.	It was necessary to declare some variables, constants, units of measure, variation orders or initial value. It offers a first graph of how the phenomenon would behave.	It was necessary to declare some variables, constants, units of measure, variation orders or initial value. It does not offer a first graph of how the phenomenon would behave.
Simulation interpretation	Interprets the simulator associated with the problem and relates it to the modelled differential equation, declaring each element and obtaining the graphs.	Interpret the simulator associated with the problem declaring each element and obtaining the graphs.	The declaration of some elements that the simulation considers is not evidenced.
Understanding of the mathematical model	It explains and understands each part of the differential equation that models the phenomenon, in addition to the initial values.	It explains the differential equation that models the phenomenon, but the initial values are not considered.	The initial values are not explained, and the explanation of the differential equation is not adequate.
Construction of the new problem	Redesign examples of applications considering the functional character of mathematical objects.	Redesigns examples of applications but of an artificial nature of mathematical objects.	Made only numerical changes to the examples of applications.

3. Results

Using the PHET interactive simulator [13] (Figure 1), students make comparisons of the pendulum's angular motion when changing body mass, friction, or string length. They also point out that there is a change in amplitude, that is, considering the comparison of possible initial values as the reference point, the body will not reach its initial position until it starts from rest. They do not verbalize the oscillating movement that the angular position has in the simulation.

They state the constants such as the mass of the body, the length of the string, friction, and gravity. As independent variable they place the time and as dependent variable they place the amplitude of the angle. Each element with its units of measure. In this qualitative study they were unable to identify angular acceleration, angular velocity, and angular position.

Figure 1. Pendulum lab offered by the University of Colorado Boulder [13].

They were able to sketch a discrete-time graph, building part of their reference frame, with respect to angle as a function of time (Figure 2). This graph identifies that they recognize this oscillating behavior of the angle from the comparison of states.

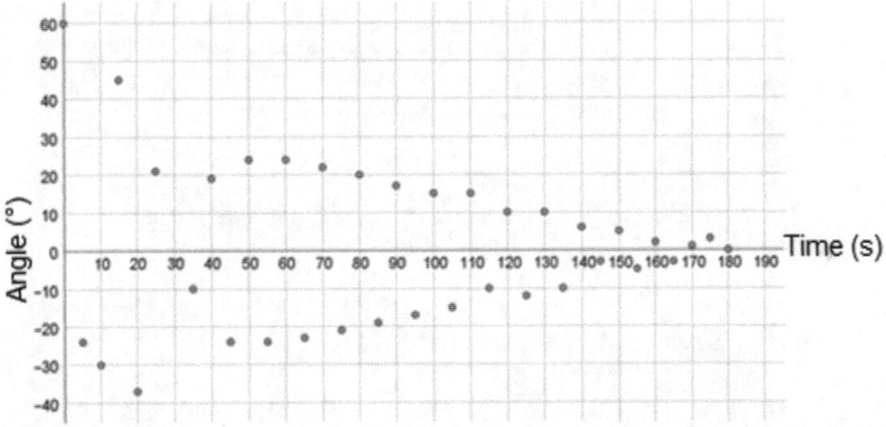

Figure 2. Graph made by the students that describes the behavior of the angle in the pendulum.

In the study of Vensim's angular position simulation (Figure 3), they must explore each of the variable, level, arrow, and rate. From the simulation they determine that the rate are changes of the variable or differences, while the level boxes are accumulations of those changes

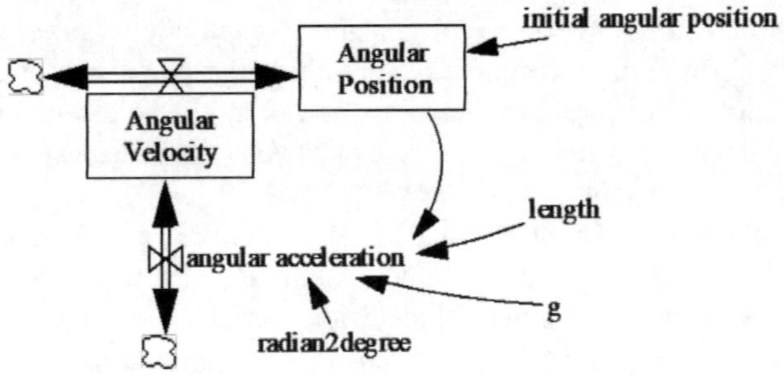

Figure 3. Vensim's angular position simulation.

Journal of Physics: Conference Series **2515** (2023) 012011 doi:10.1088/1742-6596/2515/1/012011

They confuse the value of the constants: gravity and string length; as initial values of the problem. They also mention that the initial position of the angle as well as the time at zero are initial values. They do not recognize the initial value of the angular velocity at time zero.

Regarding the change, they wrote: "As time goes by there is an increase in speed", they also mentioned: "Now, as in the change (first derivative) the speed increases, then as time goes by the acceleration increases". On the first sentence, they are only analyzing the comparison between the maximum relative points obtained in the angular position. From the second sentence the error of considering that if the derivative increases implies that the second derivative also increases.

From the graphs obtained from the simulation, they conclude a relationship between the curves: the accumulation of angular accelerations (area under the acceleration curve) determines the angular velocity, while the accumulation of angular velocity (area under the velocity curve) determines the angular position curve (Figure 4).

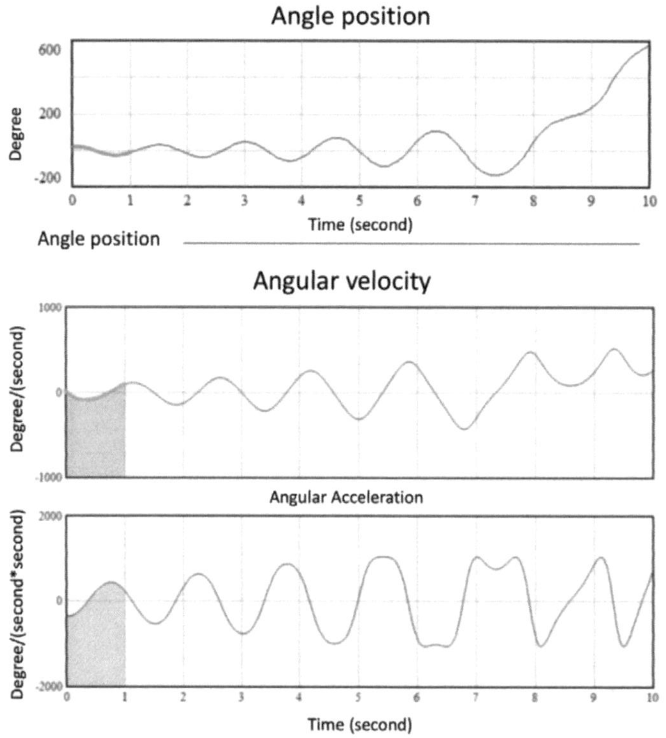

Figure 4. Students illustrated with colors the relationship between the area and the antiderivative curve.

On the other hand, they fail to interpret that the graphs obtained represent a trending movement, that is, making the angular position of the pendulum extend more and more. Defined in system dynamics as a positive feedback model. In this section they managed to adequately understand the physical results that will help to build the differential formula. They conclude that the differential equation is equivalent to the study of accumulations of differences that allow obtaining the solution of the differential formula. They also visualize that since the initial velocity is zero at zero time, then the tangent line that exists in the initial angular position is constant.

They took another problem posed as an application of the simple pendulum to carry out the modelling procedures again. It is important to point out that it was necessary to enhance the importance of the differential equation as a predictive model for decision making. The study of the accumulation of differences in values does not make functional sense if it is not used for predictive purposes.

4. Discussion

With this type of activities, some axes of formation of the mathematics education career are strengthened: reflection was made on the extra-mathematical phenomena that can be modelled, the formulation of problems by the students was promoted, strategies for the solution of a problem are discussed, technological resources that supported reflection on teaching strategies, how mathematical objects were learned or meant, on the difficulties and skills in learning the object. However, it is necessary to improve in other aspects: how to use the problems that came out of this proposal for future implementations? how to rethink the questions and moments for a better understanding of the simulation and modelling? what do the students need before using it? and understanding of simulators?

Two types of simulations are recognized: interactive digital simulation to understand the phenomenon and digital simulation using a block-oriented language. Both played an important role for the construction and understanding of mathematical objects, but both simulations have a different nature in their use. The former helps more to understand the real situational context, so that it does not stray too far from the learner's reality, while the latter helps to strengthen the meaning of the objects.

Simulation, even though it allows the model to be compared with reality, also has other advantages: it guarantees safety when working with phenomena such as electrical circuits or mixtures, the time variable does not cause problems when working hypothetically with population growth, it gives meaning to initial values when something goes wrong, saves on lab equipment, allows for illustrations of what is not physically seen as electrical charge. In addition, the simulation must be studied globally (the relationships between each element of the simulation) and locally (studying the information that each element stores).

But certain disadvantages are also observed with the way of working with the simulator: if the student plays with the interactive simulations without carrying out an order of experimentation that follows, he may lose the rationality of his actions, difficulty in understanding the Vensim language when only writing the diagram without reflecting on what has been done previously. Concerns were sometimes raised about creating a model with Vensim from scratch, a challenge to see what other strategies would use this software.

5. Conclusions

The students, when carrying out a qualitative and quantitative study of the change, were able to identify variables and constants involved in the physical system, helping to understand each term of the differential equation and the behavior of the solution. Therefore, the physical phenomenon was key for the students to signify the mathematical objects related to the second order ordinary differential equation.

Students were able to visualize the initial values (position and velocity) that helped determine a single solution curve. The students not only saw the initial values, but also interpreted them and saw their function with respect to the behavior of the curve and the construction of predictions or conjectures about the physical system. In addition, they managed to understand more easily the physical results used for the construction of the differential equation.

The study of change (qualitatively and quantitatively) contributed not only to the construction of the equation and its solution, but also to the interpretation and understanding of the physical system. In class, the students together with the teacher reflect on the number of variables or constants that were considered to study the problem, recognizing the complexity of physical systems, that with other types of considerations the result can be perfected.

Regarding the relationship with modelling and simulation, we believe that modelling allows to build a mathematical model, but the existing types of simulators help a lot to obtain them. In addition, simulation is that experimentation on a mathematical model of a system that helps compare theory with reality.

Acknowledgments
This work was supported by Centro de Investigación en Matemática y Meta-Matemática, Universidad de Costa Rica / Project # C1187.

References
[1] Arslan S 2010 Do students really understand what an ordinary differential equation is? *International Journal of Mathematical Education in Science and Technology* **41(7)** 873
[2] Dullius M M, Araujo I S, Veit E A 2011 Ensino e aprendizagem de equações diferenciais com abordagem gráfica, numérica e analítica: uma experiência em cursos de engenharia *Bolema* **24(38)** 17
[3] Hyland D, van Kampen P, Nolan B C 2018 Outcomes of a service teaching module on ODEs for physics students *International Journal of Mathematical Education in Science and Technology* **49(5)** 743
[4] Trigueros M 2014 Vínculo entre la modelación y el uso de representaciones en la comprensión de los conceptos de ecuación diferencial de primer orden y de solución *Educación Matemática* (Mexico: Grupo Santillana) p 207
[5] Czocher J 2017 How can emphasizing mathematical modeling principles benefit students in a traditionally taught differential equations course? *The Journal of Mathematical Behavior* **45** 78
[6] de Barros A M R 2007 Modelos matemáticos de equações diferenciais ordinárias aplicados à epidemiologia *Revista de Ciências Exatas e Tecnologia* **2(2)** 62
[7] Rodríguez R, Quirós S 2016 El papel de la tecnología en el proceso de modelación matemática para la enseñanza de las ecuaciones diferenciales *Revista Latinoamericana de Investigación en Matemática Educativa* **19(1)** 99
[8] Fallas-Soto R, Alfaro-Víquez H, Arias-Tencio F 2022 Formación inicial de docentes de matemática en el último decenio en Costa Rica *Fuentes de Aprendizaje e Innovacción* **3** 127
[9] Cantoral R 2020 Socioepistemology in mathematics education *Encyclopedia of Mathematics Education* ed Lerman S(New York: Springer International Publishing) p 790
[10] Cantoral R, Farfán R 1998 Pensamiento y lenguaje variacional en la introducción al análisis *Epsilon* **42(14)** 3
[11] Cantoral R, Moreno A, Caballero M 2018 Socio-epistemological research on mathematical modelling: an empirical approach to teaching and learning *ZDM* **50** 77
[12] Fallas-Soto R, Lezama J 2022 Argumentos variacionales en la comprensión de la concavidad en gráficas de funciones *Perfiles Educativos* **44(178)** 130
[13] PhET Interactive Simulations 2023 *Pendulum Lab, 1.0.24* (Boulder: University of Colorado Boulder)

Teaching mathematics in physics. a study of the mathematics programs of a Costa Rican university

C Gaete-Peralta[1]**, and R Fallas-Soto**[2]
[1] Departamento de Matemáticas y Ciencias de la Ingeniería, Universidad Bernardo O'Higgins, Santiago, Chile
[2] Departamento de Educación Matemática, Universidad de Costa Rica, Montes de Oca, Costa Rica

E-mail: claudio.gaete@ubo.cl, rodolfo.fallas@ucr.ac.cr

Abstract. The objective of this research was to describe the mathematical contents that are part of the mathematics programs of a Physics course at a certain Costa Rican university. In order to respond to the proposed research objective, data were collected from the mathematics study programs of the bachelor's degree in Physics at Universidad de Costa Rica, which were analyzed through a thematic analysis, braided with documentary analysis techniques, where the ATLAS.ti computational tool served as support to carry out such analysis. The results of this research allowed describing the mathematical contents studied in this career through ten topics. Based on these results, a discussion was held about modeling situations that can be implemented in mathematics courses of this major with the purpose of strengthening the link between mathematics and physics. Among the conclusions, the educational community is left with the task of incorporating different theoretical modeling perspectives into the mathematics programs of this career that will strengthen its link with physics.

1. Introduction

One of the reasons for teaching mathematics is to enable students to account for its use in reality [1]. However, this teaching traditionally approaches the link between mathematics and reality by means of stereotyped tasks, whose realities turn out to be artificial and contribute little to the achievement of this link [2]. In the particular case of Physics, Romero [3] states that students tend to manipulate formulas without a deep understanding of the physical concepts involved in them.

Along the same lines, mathematical modeling can be considered as a process, a strategy, or a learning environment in which the need to use, build and validate models to solve real Physics problems is involved [4]. In this way, carrying out modeling activities in mathematics courses of Physics majors is important since this type of activities contributes to the students' ability to understand physical concepts through the use of mathematics. In addition, mathematical modeling contributes to the development of competences, visions, and critical attitudes about the role of mathematics in society [5], and, in particular, in Physics.

Knowing what mathematical contents are studied in mathematical subjects in Physics majors allows knowing and dimensioning how the teaching of mathematics is linked to Physics. Furthermore, this allows to generate a discussion about modeling situations that can be implemented in mathematics courses in order to strengthen their link with Physics. With this purpose, the objective of this research is to describe the mathematical contents that are part of the mathematics programs of a Physics major at a certain Costa Rican university.

2. Methodology

The following is a description of the methodological aspects that will make it possible to respond to the research objective.

2.1. Context

The higher education institution that was part of this study was Universidad de Costa Rica (UCR), which offers a bachelor's degree in Physics. The curriculum of this program is divided into eight cycles and the mathematical subjects taught are the following: Precalculus (acronym MA-0001), Calculus I (acronym MA-1001), Calculus II (acronym MA-1002), Calculus III (acronym MA-1003), Linear Algebra (acronym MA-1004), and Differential Equations (acronym MA-1005).

2.2. Data collection

The data were collected from the study programs of the mathematics subjects taught in the bachelor's degree in Physics at UCR: Precalculus, Calculus I, Calculus II, Calculus III, Linear Algebra, and Differential Equations.

2.3. Data analysis

To analyze the collected data, a thematic analysis was used [6,7] because it is an exploratory, descriptive, and interpretative approach that allowed the identification of themes that made it possible to describe the mathematical contents that are part of the mathematics programs of the bachelor's degree in Physics at the UCR. In addition, in order to better analyze the information underlying the data obtained, the thematic analysis was combined with documentary analysis techniques [8], where such combination was carried out by experts in Mathematics Education. It is important to note that this research understands documentary analysis as a process of inference where information is thoroughly studied, interpreted, and synthesized to give rise to a formulation that underlies the original document [9]. Likewise, both the thematic and documentary analysis processes were combined with the ATLAS.ti computational tool, with the purpose of supporting and optimizing the analysis process and thus making an objective and systematic description of the information obtained from the data.

Thematic analysis consisted of six stages: (1) familiarization with the data, (2) generation of initial codes, (3) search for themes, (4) review of themes, (5) definition and naming of themes, and (6) preparation of the report. The first stage consisted of reading all the data collected from the curricula, a process that allowed a first understanding of the mathematical contents that are part of the mathematics programs studied. In the second stage, a descriptive coding was carried out to recognize certain codes of a more general nature, which made it possible to generate potential themes. It is important to note that the codes identify a characteristic of the data (semantic or latent content) that appears interesting to the analyst and refer to the most basic segment of the raw data or information that can be meaningfully evaluated in relation to the phenomenon [6]. In this sense, codes differ from topics, as the latter are usually broader [6]. A refinement of the coding process was carried out in stage 3. Stage 4 brought together the analyses of all code associations made by the experts, who worked in groups and in parallel, and subsequently agreed on the discordant codes. Stage 5 focused mainly on a refinement of the topics, where the definitions of the themes were generated to faithfully represent the data. Finally, stage 6 consisted of the construction of the topics describing the mathematical contents studied in the bachelor's degree in Physics at UCR.

It should be noted that the aforementioned groups of experts performed the first three stages of the thematic analysis separately, who subsequently discussed their results and focused on recognizing the similarities and discrepancies obtained during the process of coding and searching for themes, which made it possible to construct the topics. In other words, the data triangulation method [10] was

Journal of Physics: Conference Series **2515** (2023) 012012 doi:10.1088/1742-6596/2515/1/012012

considered for the analysis of the data obtained, which allowed minimizing the bias of a single analysis, increasing the quality and validity of the information found [11].

3. Analysis and results

During the analysis of the data, 10 topics were identified which made it possible to describe the mathematical contents that are part of the mathematics programs of the bachelor's degree in Physics at UCR. They are (1) real functions of one variable, (2) derivative of real functions of one variable, (3) integral of real functions of one variable, (4) series, (5) vector geometry, (6) functions of several variables, (7) derivatives of functions of several variables, (8) integrals of functions of several variables, (9) linear algebra, and (10) differential equations. Table 1 shows the codes obtained after data analysis.

Table 1. Codes resulting from the analysis of the data obtained from the syllabuses of the mathematics subjects.

Subject	Codes
Precalculus (MA-0001)	Real numbers; inequalities; algebraic expressions; real functions of real variable; polynomial function; rational function; radical function; absolute value function; exponential function; logarithmic function; trigonometric ratios; trigonometric identities; trigonometric functions.
Calculus I (MA-1001)	Calculus of limits; calculus of derivatives; tangent and normal lines; ratio of change; optimization; antiderivatives; integration techniques; definite integral; area calculations.
Calculus II (MA-1002)	Calculus of improper integrals; convergence criteria of improper integrals; Taylor and Maclaurin polynomials; ellipse; hyperbola; parabola; polar coordinates; graphs of polar curves; area of regions bounded by curves in polar coordinates; mathematical induction; power series; complex numbers.
Calculus III (MA-1003)	Lines and planes in $\mathbb{R}^3$; conic sections in $\mathbb{R}^3$; curves in $\mathbb{R}^3$; vectors; curvature; functions of several variables; partial derivatives; directional derivatives; optimization of functions of several variables; line integrals; surface integrals.
Linear Algebra (MA-1004)	Systems of linear equations and matrices; operations with matrices; operations with vectors; calculation of areas and volumes in $\mathbb{R}^3$; lines and planes in $\mathbb{R}^3$; vector spaces; linear transformations; eigenvalues and eigenvectors; quadratic curves.
Differential equations (MA-1005)	First order differential equations; higher order differential equations; power series; systems of first order differential equations; Laplace transform; partial derivative differential equations.

The following is an explanation of the themes encountered.

3.1. Real functions of one variable

The mathematics subjects of the bachelor's degree in Physics at UCR include the study of contents associated with real functions of one variable, where different types of functions are studied, such as polynomial, rational, radical, absolute value, exponential and logarithmic functions, along with a study of their respective graphs. Additionally, in this course we study trigonometric ratios and identities and then move on to the study of trigonometric functions and their respective graphs.

3.2. Derivative of real functions of one variable

The mathematics courses of the bachelor's degree in Physics at UCR include the study of contents associated with the concept of derivative of real functions of one variable, where limits of functions of one variable are studied and to later move on to the definition of derivative as a limit of a quotient and its respective geometric interpretation (slope of the tangent line at a point). After that, we proceed to study different derivative techniques to be used in different types of curve tracing and optimization problems.

3.3. Integral of real functions of one variable
The mathematics courses of the bachelor's degree in Physics at UCR include the study of contents associated with the concept of integral of functions of one variable, where contents such as antiderivative, definite integral, the fundamental theorem of calculus, integration techniques (simple substitution, integration by parts, partial fractions, and trigonometric substitution) and applications of the definite integral to geometric problems such as the calculation of areas are studied. Additionally, improper integrals of first and second species are studied, together with different types of convergence criteria, such as direct comparison, comparison by quotient, absolute and conditional convergence, among others.

3.4. Series
The mathematics courses of the bachelor's degree in Physics at the UCR include the study of contents associated with the concept of Series, where numerical series and power series are studied, together with different types of convergence criteria.

3.5. Vector geometry
The mathematics courses of the bachelor's degree in Physics at the UCR include the study of contents associated with vector geometry, where straight lines, planes, and curves in the space $\mathbb{R}^3$ are studied, together with the concept of curvature and torsion of a curve.

3.6. Functions of several variables
The mathematics courses of the bachelor's degree in Physics at the UCR include the study of content associated with functions of several variables, where scalar fields in two and three variables, limits and continuity are studied.

3.7. Derivatives of functions of several variables
The mathematics courses of the bachelor's degree in Physics of the UCR include the study of contents associated with the derivatives of functions of several variables, where concepts such as partial derivatives, directional derivatives, chain rule and optimization problems where the criterion of the Hessian matrix and Lagrange multipliers are used are studied.

3.8. Integrals of functions of several variables
The mathematics courses of the bachelor's degree in Physics at UCR include the study of contents associated with the integrals of functions of several variables, where concepts such as double integrals over closed and bounded regions in $\mathbb{R}^2$, changes of variables, area and volume calculations are studied. In addition, concepts such as triple integrals over cubes and closed and bounded regions in $\mathbb{R}^3$ are studied, along with various types of changes of variables. Finally, we proceed to the study of line integrals and surface integrals.

3.9. Linear algebra
The mathematics courses of the bachelor's degree in Physics at UCR include the study of contents associated with linear algebra, where matrix algebra is studied and its use in the resolution of systems of linear equations, vector spaces, linear transformations, eigenvalues and eigenvectors, and diagonalization of matrices.

3.10. Differential equations
The mathematics courses of the bachelor's degree in Physics at the UCR include the study of contents associated with differential equations, where different types of ordinary and higher order differential

equations are studied, together with their respective methods of resolution. In addition, the study of the Laplace transforms and its application in the resolution of differential equations is carried out. Finally, the concept of differential equations in partial derivatives is studied, together with some techniques for solving them.

The results of this research show that the mathematics programs of the bachelor's degree in Physics at UCR focus on the study of mathematics itself, as a scientific discipline, and do not explicitly incorporate applications in Physics or the use of mathematical modeling as a mechanism to link mathematics and Physics. For this reason, the following section will discuss modeling situations that can be implemented in mathematics courses in order to strengthen the link with Physics.

4. Discussions

The concept of mathematical modeling has become an attractive alternative to link the teaching of mathematics with Physics, acquiring increasing relevance in educational systems around the world [12]. According to Kaiser and Sriraman [13] the development of research involving the concept of mathematical modeling can be carried out from different theoretical perspectives: realist, epistemological, educational, contextual, socio-critical, and cognitive. However, there are also theoretical perspectives that understand mathematical modeling as a cultural practice that has originated the development of the so-called ethnomodeling, where it is possible to find research such as that carried out by Orey and Rosas [14], among others. All these perspectives have given the concept of mathematical modeling different conceptions, scopes, and purposes [15].

4.1. Mathematical modeling and the P-principle: some examples in physics

Within those theoretical perspectives of mathematical modeling that seek to contribute to the link between mathematics and reality, traditionally, and in accordance with Williams and Goos [16], these perspectives develop a concept of modeling that consists of a cyclical process that "links" mathematics with the real world. There, through different types of phases, a modeling process is carried out (for example, see [17]), with the purpose of emulating the mathematical knowledge of the school in people's daily life [18]. In other words, many existing modeling approaches in mathematics education usually respect a P-principle, which positions the real world and mathematics as distant units of knowledge, where there is an intention to link them by means of a cyclical process.

The following are some examples of research that make use of a mathematical modeling perspective that respects the P-principle and that contribute to the link between mathematics and Physics. From the Realistic Mathematics Education approach [19], Fernandez [20] implemented a classroom proposal to tenth grade students, which allowed characterizing and modeling periodic phenomena in Physics contexts, allowing to see the appropriation of some aspects related to the development of variational thinking around the study of periodicity in Physics contexts and the modeling of situations of periodic nature. On the other hand, Sokolowski [21] implemented a modeling activity that allowed students to carry out a scientific reasoning, in the sense of [22], to support the learning of the fundamental theorem of Calculus from a covariational reasoning.

For the development of this activity, an interactive simulator called Moving Man was used to show the relationship between the motion of a person and its respective position $x(t)$, velocity $v(t)$, and acceleration $a(t)$. This allowed students to connect the scientific foundations of the phenomenon with its mathematical representations by simultaneously observing these motion graphs. From the perspective of research-based modeling [23], Ekici and Alagoz [24] designed and implemented an activity that allowed promoting students' learning of trigonometric concepts through the modeling of periodic orbits by using GeoGebra. On the other hand, Cosmes and Montoya [25] designed and implemented a mathematical modeling activity, framed in the Mathematical Working Space theory [26], which allowed students to determine the structure deformation of a frame-type model of a statically determined structure, through the use of the concept of definite integral. Framed in a commognitive perspective [27], Rogovchenko [28] applied a mathematical modeling activity, where differential equations were used to determine the emptying time of a cylindrical pond.

4.2. Mathematical modeling and the P' principle: some examples in physics

From a socio-epistemological approach [29], mathematics is not conceived as a knowledge alien to the daily reality of the learner, but the social construction of mathematical knowledge favors the transformation of reality, and vice versa. Here, we will understand by social construction of mathematical knowledge a notion that considers the interactions and processes of debate that communities experience to carry out the processes of institutionalization of mathematical knowledge, together with its functionality in specific contexts and situations [30]. Therefore, from this approach a need arises to construct a concept of mathematical modeling that conceives a variety of the principal P, let us say P', that considers the functionality of the reciprocal relationship between mathematics and everyday life.

In this case, when we speak of variety, we refer to the idea of differentiating from the traditional conceptions of mathematical modeling in education, but without losing the unity, which, in the case of mathematics education, is identified with the idea that modeling consists of a cycle that connects the real world and mathematics [31]. Considering the P' principle allows us to stop conceiving reality as a unit of knowledge distanced from mathematics and to value people's uses of mathematical knowledge in specific situations in which they are involved.

The following are some examples of research that make use of a mathematical modeling perspective that respects the P' principle and that contribute to the link between mathematics and physics. In [32] a modeling activity related to the design of a water temperature control system for algae harvesting was designed, in which a modeling activity was implemented that allowed students to build mathematical knowledge associated with the Laplace transform. On the other hand, the research conducted by Morales, *et al.* [33] analyzed the role played by time in a socio-epistemological modeling process that sought to favor the construction of mathematical knowledge by university students. In this process, the functionality of graphs to describe a physical phenomenon associated with the movement of projectiles, using track moving objects technology, was evidenced.

Finally, Cordero [34] provides theoretical references for the design of modeling situations, from a socio-epistemological perspective, that allow students, through the use of the notion of accumulation, to generate a resignification of mathematical knowledge associated with the concept of definite integral, in different types of situations typical of Physics. Examples of these are the accumulation and circulation of fluids, electricity, magnetism, and mechanical work, where the resignification of mathematical knowledge is understood as the construction of knowledge itself, according to the organization of the human group [35].

5. Conclusions

The description of the contents that were part of the mathematics programs of the bachelor's degree in Physics at Universidad de Costa Rica allowed us to know and dimension how the mathematics taught in this degree is linked to Physics. In this sense, the results of this research allow us to conclude that these programs do not focus on explicitly linking mathematics with Physics.

Mathematics is a fundamental tool for the development of Physics. For this reason, it is important that the mathematics programs in Physics majors generate instances that allow mathematics to serve as a tool to solve Physics problems. In this sense, it remains as a task for the educational community to incorporate in the mathematics programs of the bachelor's degree in Physics of the Universidad de Costa Rica different theoretical perspectives of modeling that allow, through situations such as those presented in this research, to strengthen the link between mathematics and Physics.

Acknowledgments

Proyecto FONDECYT de iniciación nº 11220346.

References

[1] Acebo J, Rodríguez R 2019 Mathematical modeling in the educational field: a systematic literature review *Technological Ecosystems for Enhancing Multiculturality* (New York: Association for Computing Machinery) p 718

[2] Villa-Ochoa J 2015 Modelación matemática a partir de problemas de enunciados verbales: un estudio de caso con profesores de matemáticas *Magis* **8(16)** 133

[3] Romero A, Rodríguez L 2002 La formalización de los conceptos físicos. El caso de la velocidad instantánea *Educación y Pedagogía* **15(35)** 57

[4] Sánchez-Cardona J, Rendón-Mesa P, Villa-Ochoa J 2021 Proyectos de modelación matemática como estrategia de evaluación formativa en un curso para futuros profesores de matemáticas *Revista Meta: Avaliação* **13(40)** 543

[5] Villa-Ochoa J, Castrillón-Yepes A, Sánchez-Cardona J 2017 Tipos de tareas de modelación para la clase de matemáticas *Espaço Plural* **18(36)** 219

[6] Braun V, Clarke V 2006 Using thematic analysis in psychology *Qualitative Research in Psychology* **3(2)** 77

[7] Braun V, Clarke V, Hayfield N, Terry G 2019 Thematic analysis *Handbook of Research Methods in Health Social Sciences* ed Liamputtong P (Singapore: Springer) p 843

[8] Rojas I 2011 Elementos para el diseño de técnicas de investigación: Una propuesta de definiciones y procedimientos en la investigación científica *Tiempo de Educar* **12(24)** 277

[9] Cordero F, Del Valle T, Morales A 2019 Usos de la optimización de ingenieros en formación: el rol de la ingeniería mecatrónica y de la obra de Lagrange *Revista Latinoamericana de Investigación en Matemática Educativa* **22(2)** 185

[10] Arias M 2000 La triangulación metodológica: sus principios, alcances y limitaciones *Investigación y Educación en Enfermería* **18(1)** 13

[11] Gavira S, Osuna J 2015 La triangulación de datos como estrategia en investigación educativa *Revista de Medios y Educación* **47** 73

[12] Huincahue J, Borromeo-Ferri R, Mena-Lorca J 2018 El conocimiento de la modelación matemática desde la reflexión en la formación inicial de profesores de matemática *Enseñanza de las Ciencias* **36(1)** 99

[13] Kaiser G, Sriraman H 2006 A global survey of international perspectives on modelling in mathematics education *ZDM-Mathematics Education* **38(3)** 302

[14] Orey D, Rosa M 2015 Three approaches in the research field of ethnomodeling: emic (local), etic (global), and dialogical (local) *Revista Latinoamericana de Etnomodelación* **8(2)** 364

[15] Villa-Ochoa J, Rosa M, Gavarrete M 2018 Aproximaciones socioculturales a la modelación en educación matemática. Aportes de una comunidad latinoamericana *Revista Latinoamericana de Investigación en Matemática Educativa* **11(1)** 4

[16] Williams J, Goos M 2013 Modelling with mathematics and technologies *Third International Handbook of Mathematics Education* ed M Clements, A Bishop, C Keitel, J Kilpatrick, F Leung (New York: Springer) p 549

[17] Blum W 2015 Quality teaching of mathematical modelling: what do we know, what can we do *International Congress on Mathematical Education* (New York: Springer) p 73

[18] Mendoza E, Cordero F 2018 La modelación en las comunidades de conocimiento matemático. El uso de las matemáticas en ingenieros biónicos. El caso de la estabilidad *Revista Latinoamericana de Etnomatemática* **11(1)** 36

[19] Freudenthal H 1991 *Revisiting Mathematics Education* (Dordrecht: Kluwer Academic Publishers)

[20] Fernández A *Una Propuesta de Aula que Vincula la Física a la Enseñanza de la Función Seno, desde el Enfoque de la Educación Matemática Realista* (Cali: Universidad del Valle)

[21] Sokolowski A 2018 *Scientific Inquiry in Mathematics-Theory and Practice. A STEM Perspective* (Switzerland: Springer)

[22] Prince M, Felder R 2006 Inductive teaching and learning methods: definitions, comparison, and research bases *Journal of Engineering Education* **95(2)** 123

[23] Anhalt C, Cortez R 2015 Mathematical modelling: a structured process *Mathematics Teacher* **108(6)** 447

[24] Ekici C, Alagoz C 2021 Inquiry-based orbital modelling to build coherence in trigonometry *Mathematical Modelling Education in East and West* ed F Leung, G Stillman, G Kaiser, K Wong (Cham: Springer) p 477

[25] Cosmes S, Montoya E 2021 Understanding links between mathematics and engineering through mathematical modelling-the case of training civil engineers in a course of structural analysis *Mathematical Modelling Education in East and West* ed F Leung, G Stillman, G Kaiser, K Wong (Cham: Springer) p 527

[26] Kuzniak A, Tanguay D, Elia I 2016 Mathematical working space in schooling: an introduction *ZDM-Mathematics Education* **48(6)** 721

[27] Sfard A 2008 *Thinking as Communicating: Human Development, the Growth of Discourses, and Mathematizing* (Cambridge: Cambridge University Press)

[28] Rogovchenko S 2021 Mathematical modelling problems in a mathematics course for engineers: a commognitive perspective *Mathematical Modelling Education in East and West* ed F Leung, G Stillman, G Kaiser, K Wong (Cham: Springer) p 561

[29] Cantoral R 2013 *Teoría Socioepistemológica de la Matemática Educativa. Estudios sobre Construcción Social del Conocimiento* (Barcelona: Gedisa)

[30] Cordero F, Gómez K, Silva-Crocci H, Soto D 2015 *El Discurso Matemático Escolar: la Adherencia, la Exclusión y la Opacidad* (Barcelona: Gedisa)

[31] Opazo C 2020 *Identidad Disciplinar en la Formación Inicial Docente: una Resistencia al Discurso Matemático Escolar* (México: Centro de Investigación y de Estudios Avanzados del Instituto Politécnico Nacional)

[32] Mendoza-Higuera E J, Morales-Reyes J L, Giacoleti-Castillo F, Cordero F 2022 Categories of modelling and reproduction of behaviors in other disciplines: teaching mathematics in engineering *Mathematical Modelling Programs in Latin America* ed Rosa M, Cordero F, Orey D, Carranza P (Cham: Springer) p 291

[33] Morales A, Mena J, Vera F, Vera R 2012 El rol del tiempo en un proceso de modelación utilizando vídeos de experimentos físicos *Enseñanza de las Ciencias* **30(3)** 237

[34] Cordero F 2003 *Reconstrucción de Significados del Cálculo Integral. La Noción de Acumulación como una Argumentación* (México: Grupo Editorial Iberoamérica)

[35] Cordero F 2008 El uso de las gráficas en el discurso del cálculo escolar. Una visión socioepistemológica *Investigaciones Sobre Enseñanza y Aprendizaje de las Matemáticas: un Reporte Iberoamericano* ed Cantoral R, Covián O, Farfán R M, Lezama J, A Romo (México: Díaz de Santo) p 285

Journal of Physics: Conference Series 2515 (2023) 012013 doi:10.1088/1742-6596/2515/1/012013

Hydromechanical behavior of unsaturated swelling clays: review

M C Olarte[1], and A L B Cavalcante[1]
[1] Department of Civil and Environmental Engineering, University of Brasília, Brasília, Brazil

E-mail: maria.garzon@aluno.unb.br

Abstract. Expansive soils are materials composed mainly of active minerals of the smectite family, where montmorillonite is an essential member of that group. These types of soils tend to undergo volumetric changes due to variations in moisture content. Many geotechnical engineering practices, such as dam core design or subgrade compaction, require a thorough evaluation of the response of these soils to determine placement conditions. However, the main problem in evaluating the mechanical response of this type of soil lies in the nonlinear nature, which limits the use of most classical constitutive models that apply the basic rules of elastoplasticity. In this study, an analysis is carried out to understand the hydromechanical behavior of expansive soils in saturated and unsaturated conditions. Direct and indirect physical evidence is presented and discussed to evaluate the hydraulic response of the soil structural elements responsible for these adverse dynamics from a global (expansive phenomena as a whole) and individual (swelling and shrinkage) perspective. Since the response of these soils is hydrostatic dependent, the locus of water movement in porous media is examined, starting from the Richards equation (1931), in order to consider the flow mechanisms in unsaturated media under transient conditions.

1. Introduction

Expansive soils are materials composed mainly of surfactant materials of the smectite family. These types of mineral assemblages, when in contact with water, undergo volumetric variations due to changes in moisture gradients [1]. Numerous numerical and theoretical approaches focus on the study of this type of soils because the application of these materials is found in various geotechnical engineering practices such as dam cores and pavement subgrade materials. The main problem with expansive behavior is that most classical elastoplastic models cannot be applied directly since the stress and strain paths of these materials are naturally anisotropic and non-elastic (plastic). It should be understood, therefore, that expansive behavior is a global phenomenon, encompassing both swelling and shrinkage; two processes that, although they can occur consecutively, can produce isolated responses.

Pathologies in argile-materials are usually challenging to explain from the perspective of physic-mathematical theories because their problems are based on complex physicochemical processes inherent to the nature of the soil. Expansivity is an essentially intrinsic phenomenon belonging to the structure of the clay tissue but activated by external factors, such as humidity [2]. The unbalanced native electronegativity of the clay sheets requires cations' attraction, preferably from water. This process ends up swelling the interplate spaces due to the exaggerated interaction between both materials, which generally happens in illitic and montmorillonitic clay materials. Therefore, it is a coupled action from the hydromechanical point of view since water is an external actor that interacts with a structure susceptible to moisture and generates an excessive response in the material, characterized by expansion. When entering a path of desaturation, it is essential to mention that these materials can also contract, presenting an opposite, but ultimately mechanical effect that aggravates the pathology, especially when the clay is part of the foundation soil of lightweight structures [3].

Individually speaking, the swelling process is strongly dependent on electrochemistry. In a saturated condition, the bonds linking the lamellar assemblies, usually called tactoids, are weakened by the violent cation exchange, which excites the lamellar deflocculation mechanism. Most of this dynamic is dependent on mineralogy, however, on a microscopic scale, this spectrum responds to two theories: i) the diffuse double layer (DDL) and the specific surface area (SSA). The DDL is nothing more than a film of adsorbed water that forms on the outer wall of the clay sheets. This water hydrosphere preserves the electrical neutrality that is given by the cationic balance. Thus, the greater the thickness of the DDL, the greater the cation exchange of the mineral, and hence the volumetric change experienced. Unlike DDL, SSA is often used to evaluate the physicomechanical properties of materials. In the case of granular soils, the forces that dominate the mechanical behavior are mainly of Newtonian origin, such as gravity. In this type of materials, the consideration of capillary and electrochemical forces is not important, because as the grain size is larger than 0.0075 mm, the SSA is very small [4].

Clays are lamellar materials composed of aluminum silicates configured in a network that grows in complexity depending on the increase in cation exchange capacity, which is closely related to expansivity. The most expansive clays generally have a so-called 2:1 structure, i.e., a layer of octahedral alumina sandwiched between two layers of tetrahedral silica. It is now possible to estimate quantitatively and qualitatively the expansivity of clay soils. Although the problem occurs within a partially saturated soil condition, many models still consider that the pathology occurs in a saturated material situation. For this reason, it is essential to study the effect of unsaturated flow and its relation to the expansivity phenomenon, which has yet to be extensively studied. The Richards equation [5] may be necessary for its eclectic character since it condenses three crucial aspects of learning flow in porous media.

Since the discovery of the properties present in the microstructure of clay particles in the 1930s, the knowledge of the mechanisms involved in the molecular dynamics at the water/clay boundary has increased enormously in recent years. To fully understand the electrochemical interaction system and the nature of cation exchange, applying the basic theory of cation effects and the law of charge is necessary. During the saturation process, the expansion phenomenon reduces the attractive forces concerning the repulsive ones, which induces the weakening of the ionic bonds between lamellae.

Thus, once the soil/water interface is established, the tensions generated by the polarity of water easily dominate the Van der Waals forces of attraction [6], leading to swelling. The electrical or repulsive surface forces depend on the thickness of the DDL, which can increase when the concentration of exchangeable ions is reduced or when the dielectric constant is increased. Therefore, the mechanisms that directly interfere with the swelling process interact simultaneously and depend on the charge potential at the particle surface (ψ_0) and the distance from the surface (x). Figure 1 explains the swelling process using the DDL theory.

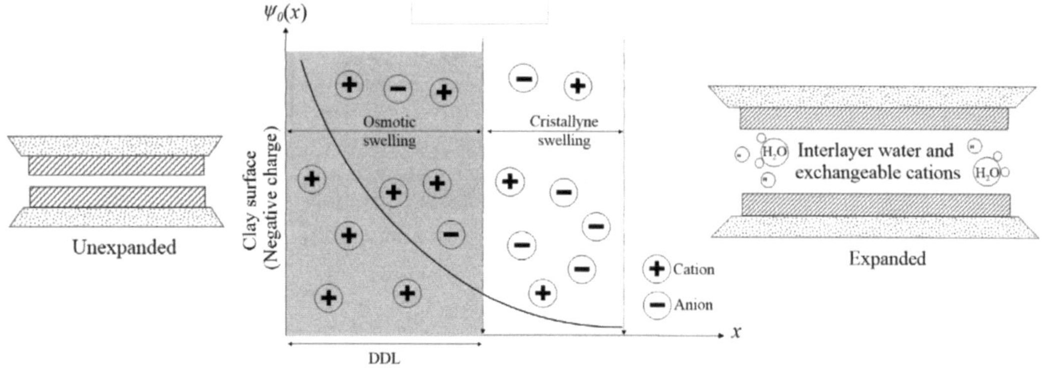

Figure 1. Schematic of swelling process.

2. Hydromechanical behavior of swelling soils

Suction (ψ) is currently investigated as one of the critical factors interfering with the hydromechanical behavior of unsaturated soils. The first analytical evaluation of unsaturated behavior was performed by Bishop, *et al.* [7] using controlled suction triaxial tests. These early constitutive approaches borrowed concepts from soil science, so the mechanical analysis was thoroughly conservative. Values of ψ can be measured or applied empirically in numerous ways, either by employing filter paper tests, triaxial equipment, or the axis translation technique [8]. From a purely theoretical perspective, the water retention curve (WRC) establishes a unique relationship between ψ and volumetric water content (θ) in porous media. For unsaturated soil mechanics, both ψ and θ are fundamental soil properties. However, ψ depends on piezometric heads and θ on the degree of saturation (Sr) and the pore and particle size distribution.

Most descriptive approaches that evaluate the WRC water-air system adjust the experimental results by employing predictive models [9]. A peculiarity in the WRC graphical framework is the presence of hysteresis, as the θ is generally more significant in the drying trajectory than in the wetting trajectory. Porous network non-uniformity, capillary condensation, and material permeability are the mechanisms responsible for this graphically observable behavior. The evaluation of the WRC's hysteretic response is vital to understanding the hydromechanical behavior of soils, either from under the stress-strain relationship [10] or by means of the principles of hydrostatics and hydraulics [11].

Within the descriptive framework of the WRC, it is possible to identify four domains: i) total saturation (θ_s), ii) funicular state, defined by Sr between 20 and 90%, iii) pendulum state, where ψ values are higher due to the formation of the capillary meniscus, and iv) residual state (θ_r). Most models used to fit the WRC attempt to reproduce the typical unimodal shape by including reference parameters such as the air entry value (AEV) corresponding to the value of ψ at which the first partially saturated voids develop. The intrinsic contour of the WRC is generally hyperbolic [12], and it is exclusively the saturation boundaries corresponding to θ_s and θ_r where it is possible to apply the effective stress principle of Terzaghi [13]. The constitutive equations generally employed to fit the data in the WRC descriptively were developed by Brooks & Corey [14], van Genutchen [12], Fredlund & Xing [15], and Cavalcante & Zornberg [16]. However, although each model employs different calibration parameters, the mathematical derivation includes flow concepts associated with the Richards equation [5]. Figure 2 presents the typical unimodal form of the WRC.

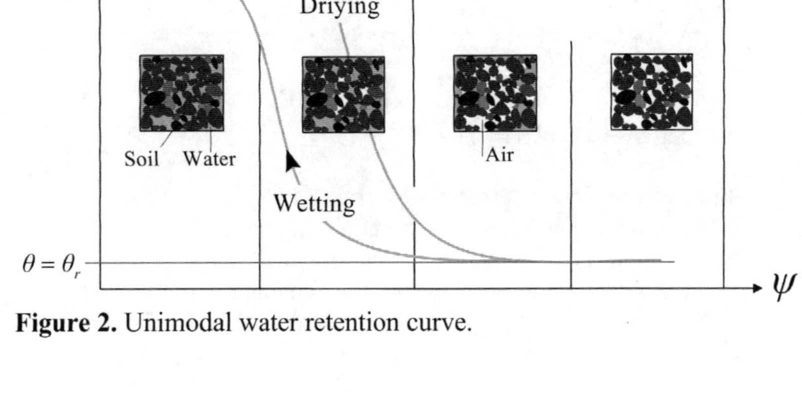

Figure 2. Unimodal water retention curve.

Under natural conditions, expansive soils can experience different stress and wetting scenarios, which can alter their hydromechanical response. There are two situations in which soils can vary their stress state: i) by loading and unloading cycles resulting from the imposition of external stresses and ii) by the lowering of the water table. On the other hand, seasonal variations induce changes in wetting fronts, leading to variations in ψ and, therefore, in shear strength. Numerous studies have investigated the hydromechanical behavior of expansive soils under stress-strain optics [17-19]. Bendahganea, *et al.* [20] employed controlled suction tests to obtain the wetting and drying trajectories of WRC and evaluate soil swelling by varying void ratio (e). Tang, *et al.* [21] conducted a campaign of edometric tests under constant θ conditions in order to study the hydromechanical response of expansive soils. However, as discussed above, the stress/strain relationship is not the only way to evaluate hydromechanical behavior. Hoffman, *et al.* [22] and Zhang, *et al.* [23] conducted infiltration tests on expansive soil samples to analyze the dependence of swelling on the hydrostatics of the soil/water boundary.

To understand the hydromechanical behavior of expansive soils, it is necessary to evaluate the mechanical spectrum from a strictly temporal perspective. During a controlled suction oedometric test, the deformation can be expressed as a percentage increase in the height of the sample (ε_{sw}) for a given time t. During the saturation process, the sample initiates the swelling stage, which process can be divided into three main modules: i) inter-void swelling (ivsw) produced by the insertion of minerals within the particular macrostructure (at this stage swelling is not significant); ii) primary swelling (psw), where about 80% of the swelling, and therefore of ε_{sw}, is produced; and iii) secondary swelling (ssw), a point reached when the relationship between ε_{sw} and t can be expressed linearly. This dynamic has been found in numerous investigations, in which the graphical archetype presented in Figure 3 is always found. In the typical unimodal t-ε_{sw} relationship, three slopes are noted, and the S-shaped curve indicates a diffusion process whose design is physically explained by Feda [24].

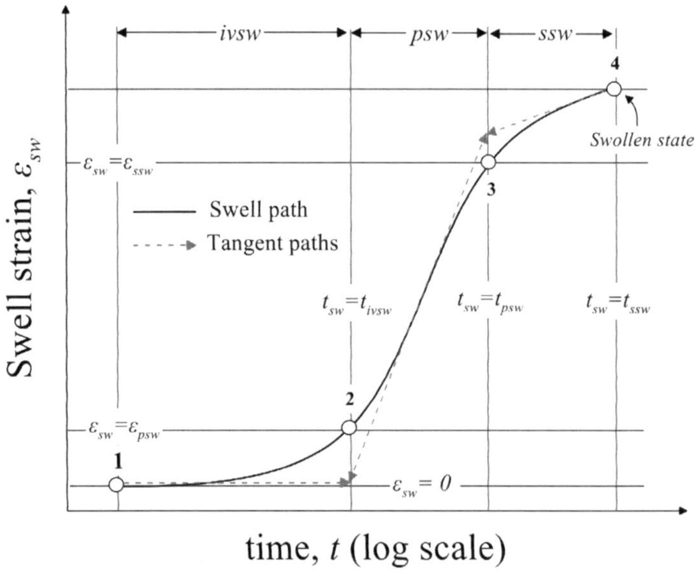

Figure 3. Swell evolvement.

3. The place of swelling soils in unsaturated soil mechanics

Due to their microscopic size, the influence of minerals on general behavior of swelling is high. Parameters such as shear strength, stiffness, loading condition and permeability make them an important control point in the swelling process. However, the hydromechanical response of swelling soils must be analyzed and addressed using the basic concepts of unsaturated soil mechanics. The first studies focused exclusively on expansive soils by Aitchinson and Holmes [25] and Norrish [26] were strongly influenced by the resources of the time, mainly derived from hydromechanical mathematics and hydrostatics. From this point on, many scientific productions were reported to broaden the knowledge of swelling and contraction processes [27-29]. The first international congress dedicated to expansive soils recognized by researchers was held at Texas A & M University in 1965. From that time to the present, a series of forums have been held annually in different parts of the world to address this soil pathology.

However, the structure of expansive clays as well as their electrochemical interaction is extremely complex, so numerous constitutive models were developed using the DDL theory, attributed to Gouy [30] and Chapman [31] in its original expression. Mathematical models employing the basic concepts of DDL do not always employ the effects derived from internal and external stresses which are widely considered factors in soil swelling behavior. Some models developed by Zhang, *et al.* [32], Phillips and Tripathy [33], and Liu [34] employ within their formulation the effects of cations and DDL operation. However, according to Mašín and Khalili [35] the above models are applicable to specific cases, *e.g.*, when the particles are suspended in low concentration monovalent electrolyte solutions.

A key aspect in understanding the place of expansive soils in unsaturated soil mechanics lies in the structure and distribution of the porous and particular spectrum in a saturated and unsaturated state. In saturated condition, the value of ψ is theoretically zero, which transfers its position to a state of "net stress" where the degree of saturation (Sr) reaches the value of 100%. Before the 1950s, expansion problems in the saturated regime were explained using Terzaghi's effective stress principle [13]. However, this principle is not fully applicable to soils that undergo volumetric variations with moisture changes since the assumptions made by Terzaghi in the original expression do not include cation exchange or the influence of electrochemical forces [36].

In the case of the unsaturated condition, most of the proposed models apply the assumptions made by Alonso, *et al.* [37] in the Barcelona Basic Model (BBM), where the relationship between ψ and Sr is inversely proportional. However, the assumptions adopted in the BBM limit the application of the classical models in the field of unsaturated swelling soils since the value of θ included in the model is much lower than the value of θ in natural condition. According to Loret, *et al.* [38], BBM-based models can estimate the swelling and shrinkage behavior in closed cases, such as in compacted bentonite samples. This is because the compaction process changes the natural pore mapping of soils, resulting in larger artificial pores and thus generating an "unsaturated" condition.

4. Conclusions

There is a global need to comprehensively evaluate the unique properties of expansive soils, not only in the point case of generally defined unsaturated soils. Rather, in the special case of volumetric changes induced by suction changes of different orders of magnitude. These volumetric variations correspond to the degree of saturation values close to 100%. Most swelling soils can remain in a saturated condition as long as the fraction of smectite on the particular surface is significant, resulting in a high-water adsorption capacity and cation hydration in the diffuse double layer. As a consequence of this electrochemical exchange process, a balance is created between the volume of adsorbed water, the volume of saturated elements, and the volume of water existing in the microporous soil structure. This means that in expansive materials that have undergone the formation of shrinkage cracks, the volume adsorbed during the swelling process is equal to the volume of air gained by the crack network.

Acknowledgments

The authors acknowledge the support of the following institutions: Conselho Nacional de Desenvolvimento Científico e Tecnológico (CNPq), Coordenação de Aperfeiçoamento de Pessoal de Nível Superior (CAPES), the University of Brasilia for funding this research.

References

[1] Nelson J, Miller D J 1997 *Expansive Soils: Problems and Practice in Foundation and Pavement Engineering* (New York: John Wiley & Sons)
[2] Olarte M C, Ruge J C, Rocha de Albuquerque P J 2021 Influence of the inclusion of synthetic compounds on the plasticity of kaolinitic clays *Arabian Journal of Geosciences* **14(16)** 1581
[3] Pérez J C, Ruge J C, Rojas J P, Orjuela J A, Cáceres J R 2021 Experimental assessment of swelling potential in montmorillonite soils *Journal of Physics: Conference Series* **1938(1)** 012006:1
[4] Santamarina J, Klein K, Wang Y H, Prencke E 2002 Specific surface: determination and relevance *Canadian Geotechnical Journal* 39**(1)** 233
[5] Richards L A 1931 Capillary conduction of liquids through porous mediums *Physics* **1(5)** 318
[6] Van der Waals J D 1873 *Over de Continuiteit van den Gas-en Vloeistoftoestand* (Leiden: Sijthoff)
[7] Bishop A W, Alphan I, Blight G E, Donald I B 1960 Factors controlling the strength of partly saturated cohesive soils *Research Conference on Shear Strength of Cohesive Soils* (New York: American Society of Civil Engineers) p 503
[8] Fredlund D G, Rahardjo H 1993 *Soil Mechanics for Unsaturated Soils* (New York: John Wiley & Sons)
[9] Cornelis W M, Khlosi M, Hartmann R, Van Meirvenne M, De Vos B 2005 Comparison of unimodal analytical expressions for the soil-water retention curve *Soil Science Society of America Journal* **69(6)** 1902
[10] Likos W J, Lu N 2004 Hysteresis of capillary stress in unsaturated granular soil *Journal of Engineering Mechanics* **130(6)** 646
[11] Parker J C, Lenhard R J 1987 A model for hysteretic constitutive relations governing multiphase flow: 1. Saturation-pressure relations *Water Resources Research* **23(12)** 2187
[12] Van Genuchten M T 1980 A closed-form equation for predicting the hydraulic conductivity of unsaturated soils *Soil Science Society of America Journal* **44(5)** 892
[13] von Terzaghi K 1923 Die berechnung der durchlassigkeitsziffer des tones aus dem verlauf der hidrodynamichen span-nungserscheinungen *Akademie der Wissenschaften in Wien, Sitzungsberichte, Mathematisch-Naturwissenschaftliche Klasse, Abteilung* **132** 125
[14] Brooks R H, Corey A T 1964 *Hydraulic Properties of Porous Media-Hydrology Papers* (Fort Collins: Colorado State University)
[15] Fredlund D G, Xing A 1994 Equations for the soil-water characteristic curve *Canadian Geotechnical Journal* **31(4)** 521
[16] Cavalcante A L B, Zornberg J G 2017 Efficient approach to solving transient unsaturated flow problems. I: Analytical solutions *International Journal of Geomechanics* **17(7)** 04017013
[17] Monroy R, Ridley A, Dineen K, Zdravkovic L 2007 The suitability of the osmotic technique for the long-term testing of partly saturated soils *Geotechnical Testing Journal* **30(3)** 220
[18] Nowamooz H, Masrouri F 2009 Density-dependent hydromechanical behaviour of a compacted expansive soil *Engineering Geology* **106(3-4)** 105
[19] Abbas M F 2016 *Multi-Dimensional Hydro-Mechanical Behavior of Compacted Highly Expansive Soils* (Riyadh: King Saud University)
[20] Bendahgane M, Nechnech A, Taibi S, Abou-Bekr N 2017 Hydromechanical behaviour of an unsaturated Algiers clay *European Journal of Environmental and Civil Engineering* **21(9)** 1093
[21] Tang C S, Tang A M, Cui Y J, Delage P, Schroeder C, Shi B 2011 A study of the hydro-mechanical behaviour of compacted crushed argillite *Engineering Geology* **118(3-4)** 93
[22] Hoffmann C, Alonso E E, Romero E 2007 Hydro-mechanical behaviour of bentonite pellet mixtures. *Physics and Chemistry of the Earth, Parts A/B/C* **32(8-14)** 832
[23] Zhang J, Niu G, Li X, Sun D A 2020 Hydro-mechanical behavior of expansive soils with different dry densities over a wide suction range *Acta Geotechnica* **15** 265
[24] Feda J 1992 *Creep of Soils: and Related Phenomena* (Amsterdam: Elsevier)
[25] Aitchinson G D, Holmes J W 1953 Aspects of swelling in the soil profile *Australian Journal of Applied Science* **4(2)** 244
[26] Norrish K 1954 The swelling of montmorillonite *Discussions of the Faraday Society* **(18)** 120
[27] Holtz W G, Gibbs H J 1956 Engineering properties of expansive clays *Transactions of the American Society of Civil Engineers* **121(1)** 641
[28] Parcher J V, Liu P C 1965 Some swelling characteristics of compacted clays *Journal of the Soil Mechanics and Foundations Division* **91(3)** 1
[29] Philip J R 1969 Hydrostatics and hydrodynamics in swelling soils *Water Resources Research* **5(5)** 1070

[30] Gouy M J 1910 Sur la constitution de la charge électrique à la surface d'un électrolyte *Journal de Physique Théorique et Appliquée* **9(1)** 457

[31] Chapman D L 1913 LI. A contribution to the theory of electrocapillarity *The London, Edinburgh, and Dublin Philosophical Magazine and Journal of Science* **25(148)** 475

[32] Zhang F, Zhang Z Z, Low P F, Roth C B 1993 The effect of temperature on the swelling of montmorillonite *Clay Minerals* **28(1)** 25

[33] Phillips A, Tripathy S 2011 Swelling pressures of some initially saturated and compacted saturated bentonites *International Journal of Advanced Technology in Civil Engineering* **1(1)** 64

[34] Liu L 2013 Prediction of swelling pressures of different types of bentonite in dilute solutions. *Colloids and Surfaces A: Physicochemical and Engineering Aspects* **434** 303

[35] Mašín D, Khalili N 2015 Swelling phenomena and effective stress in compacted expansive clays *Canadian Geotechnical Journal* **53(1)** 134

[36] Lambe T W, Whitman R V 1959 *The Role of Effective Stress in the Behavior of Expansive Soils* (Cambridge: Soil Engineering Division, Massachusetts Institute of Technology)

[37] Alonso E E, Gens A, Josa A 1990 A constitutive model for partially saturated soils *Géotechnique* **40(3)** 405

[38] Loret B, Hueckel T, Gajo A 2002 Chemo-mechanical coupling in saturated porous media: elastic–plastic behaviour of homoionic expansive clays *International Journal of Solids and Structures* **39(10)** 2773

Improvement of maintenance and asset management strategy in a maritime terminal under regulatory and physicochemical testing criteria

E A Arrieta-Estrella[1], **and L M Dueñas-Ramírez**[1,2,3]
[1] Grupo de Investigación en Estudios de Manteamiento, Universidad EAFIT, Medellín, Colombia
[2] Grupo Innovación e Ingeniería, Corporación Universitaria Minuto de Dios, Bello, Colombia
[3] Materials Science and Technology Research Group, Foundation of Researchers in Science and Technology of Materials, Bucaramanga, Colombia

E-mail: lmduenasr@eafit.edu.co, eaarrietae@eafit.edu.co

Abstract. This work was developed in the gas and oil sector, more specifically in the equipment that make up the Coveñas maritime terminal, one of the main oil terminals in Colombia. This work sought to improve the maintenance strategy and asset management of all terminal equipment through criteria based on international standards such as International Organization for Standardization 14224, which guides the ranking of equipment; also considered were the results of physicochemical tests that were applied to sample materials obtained from the equipment in the field. With this information, a road map was drawn up to establish a maintenance and asset management plan to improve the efficiency of the marine terminal's operation. The physical variables measured in this equipment and its parts, evidenced the importance of carrying out projects with interdisciplinary approaches, to enhance the achievement of the objectives set. In this specific case, a considerable improvement in maintenance work execution hours was shown.

1. Introduction

Given its strategic and functional geographic condition within the maritime operations of the hydrocarbon sector, the Coveñas maritime terminal Colombia, which is a strategic point for the export operations of crude oil from Colombia and import of refined derivatives for internal consumption. This terminal is made up of 6 asset segments, in which it is of vital importance to give specialized attention of integral maintenance; to comply with high levels of reliability during the margins or windows of operations established for the reception of oil tanker vessels [1].

These asset segmentations are classified as follows: first tanker load unit (TLU); second TLU; subsea pipelines; floating and subsea flexible hoses; service jetty; auxiliary buoys and smaller vessels [2]. The maintenance strategies of these systems are governed mostly by international standards established by highly recognized regulatory institutions, such as: The Oil Companies International Marine Forum (OCIMF), International Association of Classification Societies (IACS), American Petroleum Institute (API) and recommended practices of the National Association of Corrosion Engineers (NACE) for cathodic protection and corrosion risk mitigation of these assets; in turn complemented with strategies of the tactical and operational pillar of maintenance, such as risk based inspection (RBI), reliability centered maintenance (RCM), and condition based maintenance (CBM).

The taxonomy and hierarchy of the equipment classification tree of the systems at the Coveñas maritime terminal is structured under the International Organization for Standardization (ISO) 14224 standard [3] in its levels 6 to 9, in which the six (6) main systems were categorized, described above with their subunits and associated components for the generation of the integral maintenance strategy; which is associated to the CMMS-SAP [4], system with its time, cost and activity task lists applied within the preventive, predictive, condition-based and mechanical integrity routines plan [3].

It was determined the need to adjust the roadmaps applied to the maintenance plans of the equipment tree and to better structure the capex and Operational Expenditures (OPEX) investment costs of the integral maintenance project of the offshore infrastructure of the terminal, considering the requirements of the international regulations OCIMF, International Maritime Organization (IMO) and Dirección General Marítima (DIMAR) of Colombian government, in order to meet a high standard of operation and safety of the terminal [5].

Based on the above, the review and improvement of the previous maintenance plan from 1982 to 2012, derived from the construction of the submarine pipeline system and installation of tank units for export, which remained in force at the Coveñas maritime terminal, Colombia, until 2021, was carried out, which allowed establishing a differentiator of major maintenance activities based on regulatory recommendations and high investment costs according to the risk matrix; These were catalogued as Capital Expenditure (CAPEX) and those of routine preventive investment with the specialties mechanical, electrical, instrumentation and controls, and mechanical integrity of static equipment, as OPEX; which led to the formulation of an annual investment plan refined in man-hour costs and with a better perspective for the terminal's partner subsidiaries.

2. Methodology and materials

The methodology used in the development of this project is synthesized from the methodologies presented in [6,7] and establishes the following steps.

- Perform a comprehensive review of the routings applied to planned and scheduled maintenance work orders.
- Establish a methodology guidance document for structuring task lists and verification of maintenance times.
- Review the mechanical integrity of the assets through physicochemical tests.
- Establish efficiency and effectiveness indicators for the implementation of the routines.
- Create mechanical integrity plan and condition-based actions.

3. Results and discussions

In compliance with the first point of the methodology, a suitable classification of families, classes, and types of categories in the assets was generated, for which templates and documents were included within the quality management system, which were used as a methodology to feed the systematized database of maintenance management; following the guidelines of its central matrix in all its upstream, midstream and downstream business units.

Given the above, the integral coordination of maintenance generated a structured review agreement of the different strategies and maintenance methodology in the Coveñas maritime terminal, Colombia; Therefore, it carried out in-depth reviews of reliability indicators, maintainability and availability of transportation and storage systems in area (operational condition of crude oil storage tanks and pumping stations to offshore pipelines) and review of the mechanical integrity through the different physicochemical analysis among which stand out the micro hardness test which is a method that measures the resistance of a material to plastic deformation or penetration of an indenter Vickers hardness (HV) is calculated by optically measuring the diagonal lengths of the impression left by the indenter [8].

There is also the corrosion reaction analysis which can be defined as a metal ionization reaction, referring to when an atom loses electrons and consequently obtains an excess of positive charges, the positive ion is the cation, and the negative ion is the anion [8]. If M is the metal and of valence n, the electrochemical reaction is defined by Equation (1) defined in [8].

$$M \rightarrow M^{+n} + ne^{-}, \tag{1}$$

where M^{+n} is its ionic form and n the number of electrons yielded. Figure 1 shows the energetic changes in the corrosion process [9].

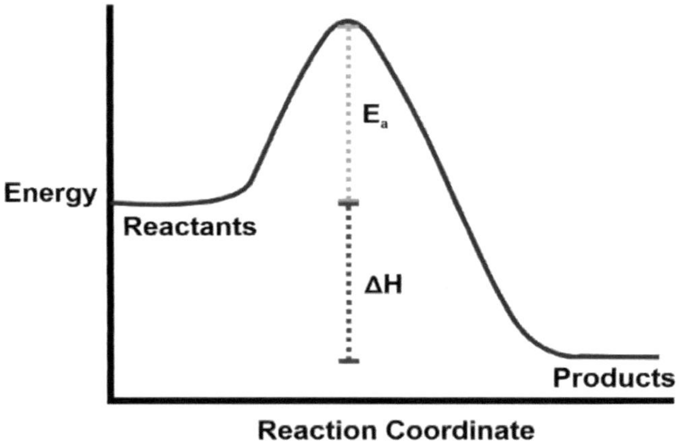

Figure 1. Energetic changes in the corrosion process [9].

The Equation (2) was used to calculate the corrosion rate (CR) [10].

$$CR = \frac{\text{Weight loss (g)} \times K}{\text{Alloy density}\left(\frac{g}{cm^3}\right) \times \text{Exposed area } (cm^2) \times \text{Esposure time (hr)}}, \tag{2}$$

where K, is a corrosion rate constant, called K Factor; this constant takes a value of K is 8.75×10^4; to finish establishing a mechanical integrity plan and condition based actions applicable to the offshore infrastructure based on the risk and safety matrix of organizational processes, leveraged under OCIMF and API normative parameters; complemented with RCM, RBI strategies and determining the corrosion rate and residual life according to API 510 [12] and API-570 [13]; which indicates that the corrosion rate will be found using equation 1 and for residual life equation 2 will be used; considering t as thicknesses [14].

Given the adjustments previously indicated and with the implementation of the maintenance plan, aligned with the quarterly programming agreements established with the asset management department, the key indicators for the organization were monitored, which include maintenance costs, backlog of work orders, effectiveness of execution, man-hours worked, availability of systems and conditions affecting the operational windows for loading vessels; for major maintenance interventions to the infrastructure; for changes of flexible hoses, subsea valves and scheduling of periodic inspections in operational conditions of tanker units. As well as conservation of mechanical integrity with coating applications on surfaces exposed to marine environment and natural deterioration by the environment; all the above with follow-up in CMMS platform for generation of monthly reports, required for presentation in follow-up meetings with partners of the Coveñas maritime terminal, Colombia.

Derived from these programming agreements, a mechanical integrity plan was established for the offshore infrastructure, which was chosen to be implemented as a strategy for major maintenance and CBM of static equipment (pipelines and TLUs), in which follow-up was established by remotely operated vehicle (ROV) robotic underwater inspection techniques, civil stabilization of the right of way of the underwater pipelines, follow-up of cathodic protection and bathymetry of the seabed, which are key activities for generating information to maintain the reliability and certification required by OCIMF.

After performing the tasks described above, the hierarchical structure and taxonomy of the available assets was verified, including the inventories of spare parts or critical stock elements in the terminal called workshop space, which resulted in an adjustment in the short-, medium- and long-term needs, as well as the disincorporation of assets, according to the criticality assessment of the terminal assets. In addition, we continue with the good practice of making maintenance programming agreements on a quarterly basis, where the key indicators established by the organization are measured and compared with the company's different dash boards at the national level.

From the previous review and improvement exercise of the hierarchical and taxonomic classification of assets established in the ISO 14224 standard [3] and with the data obtained from the physicochemical tests of micro-hardness, corrosion, among others; in addition to internal documents of the quality management system of the operating and maintenance subsidiary of the terminal, we can graphically represent the comparative of the most representative indicator in contrast of the initial strategy vs. the improved active maintenance plan of the Coveñas maritime terminal, Colombia. The man hours (MH) parameters of the initial maintenance plan strategy are shown in Table 1.

Table 2 shows the result of the application proposed in this work where the roadmaps and maintenance plans were structured in a better way, which allowed to better focus the maintenance activities to the different systems that make up the assets of the Coveñas maritime terminal, Colombia, from which the following dedication to the annualized plan to be implemented by the new operating and maintenance subsidiary of the maritime terminal was obtained.

We also note that for the TLU systems, there is a minimal percentage difference in man-hours dedicated; this was due to an additional auxiliary system in one of the TLU's that is not available in one of them; therefore, there is no Pvo review routine that applies to it. It can also be stated that the number of man-hours dedicated to these tasks was reduced by 1212 hours. This considerable reduction in man-hours is due to the combination of maintenance strategies and physicochemical tests to establish the real state of the organization's assets.

Table 1. Percentage of man-hours dedication per system of the old plan.

		Planned man-hours in current annual plan in SAP	% Plan dedication
Item	General systems	Man-hours in maintenance	%
1	Tlu1 monobuoy systems	1209	24%
2	Tlu3 monobuoy system	1571	31%
3	Subsea line system	1345	26%
4	Floating and subsea hose systems	498	10%
5	Spring system	127	2%
6	Auxiliary buoys and contingency vessels	362	7%
	Total man-hours	5112	100%

Table 2. Percentage of man-hours dedication per system of the new plan.

		Planned man-hours in current annual plan in SAP	% Plan dedication
Item	General systems	Man-hours in maintenance	%
1	Tlu1 monobuoy systems	820	21%
2	Tlu3 monobuoy system	805	21%
3	Subsea line system	504	13%
4	Floating and subsea hose systems	380	10%
5	Spring system	258	7%
6	Auxiliary buoys and contingency vessels	1133	29%
	Total man-hours	3900	100%

4. Conclusions

The present work exposes the importance of joining technical regulations, the experience of an organization, physicochemical tests and worldwide known strategies in maintenance and production to seek a significant improvement in the maintenance plan of an organization. At this point we specifically want to highlight the importance of measuring the physical and mechanical variables, to establish the real state of the assets to be managed at the Coveñas maritime terminal, Colombia. The results obtained were able to establish the necessary times to fully comply with regulatory requirements without affecting the financial budget of the maintenance strategy drastically. On the contrary, a significant improvement in the efficiency of this process was obtained.

References

[1] Dirección General Marítima (DIMAR)2018 *País de Mares No. 6* (Bogotá: Dirección General Marítima)
[2] Camerano Ruiz R E 2019 *Guía para Identificar Equipos Críticos por Seguridad de Procesos* (Colombia: Ecopetrol S.A.S.)
[3] International Organization for Standardization (ISO) 2016 *Petroleum, Petrochemical and Natural Gas Industries-Collection and Exchange of Reliability and Maintenance Data for Equipment, ISO 14224* (Switzerland: International Organization for Standardization)
[4] System Analysis Program Development 2022 *SAP, S/4HANA Cloud* (Walldorf: SAP)
[5] Lizcano Ferrer J F 2021 *Procedimiento para la Gestión Integral del Mantenimiento y Confiabilidad de Activos sobre El ERP SAP PM* (Colombia: Cenit)
[6] Jiménez Arango C A, Dueñas-Ramírez L M, Castaño Restrepo C A 2021 Optimization of maintenance management indicators in a chilled water plant: case study in the education sector *Cases on Optimizing the Asset Management Process* ed Gonzalez-Prida V, Crespo-Marquéz A (Hershey: IGI Global) Chapter 12 p 302
[7] Dueñas-Ramírez L M, Villegas-López G A, Chamat-Torres C P 2020 Application of virtual technologies for the representation of physical phenomena in maintenance tasks *Journal of Physics: Conference Series* **1587(1)** 012023:1
[8] Santos Ramos I 2021 *Evaluación de la Microdureza y Resistencia a la Corrosión de Aleaciones Nanoestructuradas Al-Mg-NTCs y Propiedades Catalíticas en NTCs y NTCs-NPsAg* (Morelia: Universidad Michoacana de San Nicolás de Hidalgo)
[9] Velez-Sanchez H 2013 *Recubrimientos de 560 AS Producidos con el Sistema de Proyección Térmica por Arco* (Bogotá: Universidad Nacional de Colombia)
[10] Suegama P H, Fugivara C S, Benedetti A V, Fernández J, Delgado J, Guilemany J N 2005 Electrochemical behavior of thermally sprayed stainless steel coatings in 3.4% NaCl *Corrosion Science* **47(3)** 605
[11] Mohan D G, Gopi S 2021 Influence of In-situ induction heated friction stir welding on tensile, microhardness, corrosion resistance and microstructural properties of martensitic steel *Engineering Research Express* **3(2)** 025023:1
[12] Amerian Petroleum Institute (API) *Pressure Vessel Inspection Code: In-Service Inspection, Rating, Repair, and Alteration, API 510* (Houston: Amerian Petroleum Institute)
[13] Amerian Petroleum Institute (API) *Piping Inspection Code: In-service Inspection, Rating, Repair, and Alteration of Piping Systems, API 570* (Houston: Amerian Petroleum Institute)
[14] Reyes Angarita J J, Vargas Jerez M, Parra C 2019 *Análisis de Costo y Ciclo de Vida para la Reposición de un Motor de Gas Superior 16G825 y Compresor de Gas Worthington tipo OF6XH6 por un Motor a Gas Waukesha 7044 S5 - Compresor Superior MH64* (Colombia: ASME Elite Training)

AUTHOR INDEX